The Gentleman, the Virtuoso, the Inquirer

The Gentleman, the Virtuoso, the Inquirer: Vincencio Juan de Lastanosa and the Art of Collecting in Early Modern Spain

Edited by

Mar Rey-Bueno and Miguel López-Pérez

Cambridge Scholars Publishing

The Gentleman, the Virtuoso, the Inquirer: Vincencio Juan de Lastanosa and the Art of Collecting in Early Modern Spain, Edited by Mar Rey-Bueno and Miguel López-Pérez

This book first published 2008 by

Cambridge Scholars Publishing

12 Back Chapman Street, Newcastle upon Tyne, NE6 2XX, UK

British Library Cataloguing in Publication Data
A catalogue record for this book is available from the British Library

ISBN (10): 1-84718-648-3, ISBN (13): 9781847186485

TABLE OF CONTENTS

INTRODUCTION

LASTANOSA, ART AND SCIENCE IN BAROQUE. A HISTORIOGRAPHICAL EXPERIMENT

MAR REY-BUENO & MIGUEL LÓPEZ-PÉREZ

In 1645 Vincencio Juan de Lastanosa published in Huesca, a northern spanish city, his first book, the *Museo de las medallas desconocidas españolas*, a detailed study of more than a hundred seventy ancient coins. Lastanosa (1607-1681) was the owner of this collection. Patron, collector, scholar and editor, Lastanosa was born in a family of merchants, priests and local rulers[1]. Comfortably installed in his palace, Lastanosa dedicated all his life on collecting marvels and curiosities, antiquities and books, studies and knowledge. Traditionally known as the patron of Baltasar Gracián, Lastanosa has being considered by himself only in the last years. This interest is promoted by the Instituto de Estudios Altoaragoneses (IEA), a local institution devoted to Huesca and its history. In 2007 the IEA celebrated Vincencio Juan de Lastanosa fourth centenary birth with several events, one of them was the International Conference *Lastanosa. Art and Science in Baroque*, origin of this book.

***

In 2005 Miguel López was invited to participate in the so-called *Lastanosa Project*, an ambitious program devoted to the study of Vincencio Juan de Lastanosa, his collections and the city where he lived, Huesca. Some years before, López had published an approach to Lastanosa chemical practices[2]. The IEA was interested in this research because offered a new Lastanosa

[1] JOSÉ IGNACIO GÓMEZ ZORRAQUINO (2004), *Todo empezó bien. La familia del prócer Vincencio Juan de Lastanosa (siglos XVI-XVII)*, Zaragoza, Diputación Provincial.
[2] MIGUÉL LÓPEZ-PÉREZ (2003), *Asclepio Renovado. Alquimia y Medicina en la España Moderna (1500-1700)*, Madrid, Ed. Corona Borealis, pp. 268-284.

face. Six months later, the IEA decided to celebrate an International Conference, devoted to splead Lastanosa beyond the aragonese frontiers.

In *Lastanosa. Art and Science in Baroque*, coordinated by Mar Rey-Bueno and Miguel López-Pérez, we explored the history of Lastanosa as scientific collector. The Conference was articulated in five sections, correspondent to some other highlighted aspects in Lastanosa universe: Collection, Garden, Laboratory, Library and Cabinet. *Collection* took as aim the study of Lastanosa cabinet of curiosities. *Garden* was devoted to the knowledge of the gardens created in the enviroment of his palace. *Laboratory* was destined to discover the unknown facet of a Lastanosa collector of secrets and chemical practitioner. *Library* analyzed the books, manuscripts, maps and other curiosities collected by Lastanosa. *Cabinet*, finally, appeared as a metaphor of the circle of writers, artists and intellectuals who had relation with Lastanosa.

***

The working guideline followed in *Lastanosa. Art and Science in the Baroque* have come out the habitual dynamics of these meetings. Even it might be said that we are before an innovative experiment. The participants did not know nothing related to Lastanosa. This is why the Organizing Committee was deciding to facilitate them copies of the principal sources that remain related to the Lastanosa collections. They are the so-called *Descripción en prosa*, *Descripción en verso*, *Catálogo Sparvenfeldt, Narración de 1662* and *Habitación de las Musas*.

Descripción en prosa was written about 1650 by Juan Francisco Andrés of Uztarroz, Chronicler of the Kingdom of Aragon and personal friend of Lastanosa. Entitled *Descripción de las Antigüedades y Jardines de Don Vincencio Juan de Lastanosa*, it is the most detailed and important description on the collections of Lastanosa. We know an only copy, in the Hispanic Society of America, which has been the used one[3].

Descripción en verso is a poetical version of the previous one, written by Uztarroz in 1647 and printed in Saragossa with the title of *Descripcion de las antiguedades, i jardines de Don Vincencio Iuan de Lastanosa ... /*

[3] Ms B2424, ff. 24r-51v°.

dscriuiala [sic] El Solitario al Dor Don RFancisco [sic] Filhol... We have selected the copy preserved in the Biblioteca del Palacio Real de Madrid[4].

Catálogo Sparvenfeldt refers to the manuscript catalogue that Lastanosa did of the books and curiosities preserved in his library. Entitled *Catálogo de los libros de D. Vincencio Ioan Lastanosa. Por orden de alfabeto*, is an exhaustive record of books, manuscripts, maps, scientific instruments, coins and antiquities hoarded by Lastanosa along his life. Two copies of the Catalogue existed. One is an eighteenth century extract. The other one is a complet copy preserved in the Royal Library of Stockholm from the end of seventeenth century. Everything relative to books and manuscripts was published by Karl Ludwig Selig in 1960. Nevertheless, Selig did not include in his edition the part of the catalogue in which Lastanosa detailed the maps, mathematical instruments, coins and antiquities of his collection. That is why we appealed the original manuscript as source[5].

Narración de 1662 is a manuscript relation that Lastanosa did of his collections. Under the title *Narración de lo que le pasó a Don Vincencio Lastanosa a 15 de octubre del año 1662 con un religioso docto y grave*, remain two copies. One, incomplete, in the Biblioteca Nacional de Madrid[6]; and the other one, preserved in the Hispanic Society of America, which is the used one[7].

Habitación de las Musas is a manuscript praise that the son of Lastanosa, Vicente Antonio, wrote of the collections of his father. Entitled *Habitación de las musas, recreo de los doctos, asilo de los virtuosos. Elogio de Lastanosa escrito durante los últimos años de su vida por su hijo y heredero Vicente Antonio* remains just an exemplar in the Hispanic Society of America, which is the used one[8].

***

The selection of these primary sources differentiates our International Conference from everything made till now. Previous the existence of Lastanosa Porject the historians who studied the life and works of the Aragonese patron were using two basic sources. On one hand, Adolphe

---

[4] sig. IX/5024(2).
[5] Ms U-379.
[6] Ms. 18727[55].
[7] Ms B2424, ff. 52r-79v°.
[8] Ms B2424, ff. 1r-5v°.

Coster article "Une description inédite de la demeure de Don Vincencio Juan de Lastanosa" (*Revue Hispanique*, 1912, XXVI, pp. 566-610), where is reproduced a seventeenth century description of Lastanosa palace, gardens and collections, entitled *Las tres cosas más singulares que tiene la casa de Lastanosa en este año de 1639*[9]. On the other hand, Ricardo del Arco book *La erudición aragonesa del siglo XVII en torno a Lastanosa* (Madrid, 1934), where is reproduced the totality of primary sources preserved in the Spanish libraries. The Organizing Committe don't use these works and the reason was simply: we selected only the reports and chronicles made by Lastanosa or his contemporaries. Del Arco selection included copies realized in eighteenth and nineteenth centuries, defectives and incompletes and *Las tres cosas mas singulares...* is a falsification, as some historians have discovered recently[10].

***

There is a last novel aspect of *Lastanosa: Art and Science in the Baroque*. The approaches made in last twenty years saw Lastanosa as an artistic and literary patron[11]. Our approximation has been done from the scientific perspective or, as would say Lastanosa contemporaries, considering him as a natural philosopher.

***

Lastanosa: Art and Science in the Baroque was an experiment. In *The Gentleman, the Virtuoso, the Inquirer. Vincencio Juan de Lastanosa and the Art of Collecting in Early Modern Spain* we can see the results. Amazing results, we can conclude. We wish to acknowledge the professional assistance of all IEA members, especially Fernando Alvira, Pilar Alcalde, Carlos Garcés and José María Esparza. Our sincere gratitude is due to Harold J. Cook, Antonio Barrera, John Slater, Jorge Cañizares-

---

[9] Biblioteca Nacional de Madrid, Ms. 18727[45].

[10] FERMÍN GIL ENCABO (1999), "La ficción *telamoniana* de Pellicer en torno a Lastanosa", en: *Actas del V Congreso de la Asociación Internacional Siglo de Oro*, Madrid, pp. 623-634 y FERMIN GIL ENCABO (2003), "Lastanosa y Gracián: en torno a Salastano", en: *Actas del I Congreso Internacional Baltasar Gracián: pensamiento y erudición*, Huesca, I, pp. 19-60.

[11] *Signos. Arte y Cultura en Huesca: De Forment a Lastanosa. Siglos XVI-XVII*, Huesca, Diputación Provincial, 1994 y JOSÉ ENRIQUE GIL LAPLANA (ed.), *La cultura del Barroco. Los jardines: arquitectura, simbolismo y literatura (Actas del I y II Curso en torno a Lastanosa)*, Huesca, IEA, 2000.

Esguerra, Daniela Bleichmar, María M. Portuondo, William Eamon, Bruce T. Moran, Rafael Chabrán and Anne Goldgar, who accepted to participate in this book. Finally, our thanks to the city of Huesca for its hospitality during a wonderful week.

CHAPTER ONE

LASTANOSA AS AN EXAMPLE OF HIS TIME: NATURAL HISTORY AND MEDICINE

HAROLD J. COOK

Is it possible to summarize what has been uncovered and reported about Lastanosa in the past few years? It may not even be possible to sum up a life of someone we have known well in person, much less someone from long ago who appears to us only through a few remains from his material life. Some of those remains have preserved the traces of his hand in the words he wrote on paper, and the words in turn preserve traces of what people like to call his thoughts. Most authors are tempted to look into the crystal ball provided by these pieces of evidence and have tried to see a person in it. But although it is tempting to think that in that glass globe we are seeing shadows of his life, what each of us sees may only be misshapen reflections of our own worlds. It is therefore sometimes preferable to compare the indications of Lastanosa's life to the remaining traces of other persons of his time who engaged in activities similar to his. In this way, Lastanosa becomes an example of the world of which he was a part. Or to put it in terms he himself might have used, he is a microcosm in which we can see the macrocosm. Of course this approach may simply be wrong: he may have been a completely unique individual who lived his own life as if only his life mattered, not being a very good example of anything very much. But by making comparisons between himself and others, and noting the differences, we can glimpse a few indications of the particular possibilities and limitations of his own situated life and even, perhaps, gain some sense of his conscious aims and intentions.

The possibilities for comparing Lastanosa to other figures in his period have arisen because of the very special nature of the conference from which the papers in this volume are drawn. The conference in turn depended very much on two kinds of historical activity: some remarkable recent detective work on Lastanosa, and the connection of Lastanosa to the history of science and ideas elsewhere in Europe. Celia Fonatana Calvo, Francisco Páez de la Cadena, and especially Carlos Garcés Manau scoured the archives and collections of material culture for anything connected to Lastanosa, and compared maps and drawings from his time to what remains in surviving buildings and on the ground today. With the generous support of the local cultural foundation, Miguel López and Mar Rey Bueno were also able to bring together the experts on Lastanosa and Huesca with two other groups of historians who have studied the period in which Lastanosa lived, those who work on the culture of Spain and on the history of science and related subjects. All were sent the same group of documents from or about Lastanosa, and asked to read them and to think about their implications. For many of the participants in the conference, then, Lastanosa was well-known as an important local figure of the 17th century, while for others of us he was hardly known until we were asked to consider him for the occasion. The result was quite unique: because we all had a common set of documents to think about, but were approaching them from many different points of view, we had not only multiple perspectives but many questions in common. Those of us who were just introduced to Lastanosa wondered about him and about Huesca and Aragon during his time, while those who knew much about his local circumstances had questions about his possible relationships to other people and places. Consequently, over the course of the days of the conference, not only when presenting papers and discussing them but when touring the city and even eating together, we had sometimes very intense debates. Who was Lastanosa, really? What was he trying to do? And what does his example tell us about intellectual life in 17th century Aragon, and Iberian Spain more generally?

Perhaps one can begin with the physical evidence. Several of the authors address questions such as: What did his house look like? How was his garden laid out? Where was his library located? What were the books in his library? The recent research presented in this volume has communicated a great deal about such matters, which has been very helpful in giving us a sense of who Lastanosa was. All of those who have examined the physical evidence have also indicated that there is much more to discover. Partly this is because documentation is incomplete, but partly because the documentation cannot be taken at face value, either when it is in the form of an image or in the form of lists in the archives. We know that some things were left out of the record, and others have not yet been searched for. Moreover, if we start thinking of the missing items, we must begin to consider matters like the relationship of his house and garden to other ones in Huesca, including the Jesuit school across the road from his front door. And what was in the other libraries and gardens of the city compared to Lastanosa's? Or compared to others in Aragon, or in northern Spain more generally? There are even puzzles about the sources of his wealth, which made his estate and collections possible. Such considerations also lead one to wonder how he treated his family, servants, and employees. Questions about these aspects of his personal life lead one to ask what his views on the political and religious controversies of his day might have been, and who were his friends, who his enemies?

Such questions may seem to be far from the focus of understanding his investigations into the natural world, but comparisons might help us to understand something more about Lastanosa's motivations and his achievements, giving them due proportion. My own impression from what has been found is that given the size of his house and garden, and its placement at the edge of the city--and given the character of the Lastanosa tomb and the two chapels he paid for--the Lastanosa family was indeed a very important one in Huesca, but perhaps not the most powerful or influential one. Were his activities a form of display that were intended to edge him a little further up the social ladder? Or were they intended to demonstrate his associations with a particular group, or party, in Aragon, or in Spain more generally? His military service in the war against Catalonia, at a time when the Aragonese had resentments of their own against Castile, might have been something into which he put his heart, or it might have been something he did because he hads little choice. Does his cultural activity help to place him on the side of the Jesuits and the court in Madrid, or suggest that he advocated cosmopolitanism? What would the other alternatives have been in his city in the middle of the 17th century? Lastanosa appears to have spent his money on his art and science, rather than intending to make money at it, as an apothecary might. It caused him briefly to achieve fame, and even to make Huesca better known to the larger world: almost four centuries later, his name still brings visitors and scholars to his home town. But only further investigations into Huesca's history will enable us to compare properly his life in general to others of his day, and so to more fully assess the personal, religious, and political aims of his projects.

Yet in the context of the main subject of the conference, Lastanosa's relationship to "art and science in the Baroque", another set of questions has provided more illumination. Except for some comments on Lastanosa' garden designs and the paper by Jorge Cañizares-Esguerra, few authors considered the importance to Lastanosa of what most people today think of when using the word "art": fine art. But at the beginning of the conference, Aurora Egido provided some very important clues for considering the relationship between art and science more generally in Lastanosa's time. In writing of the *Criticon* she noted that in the common understanding of the period "art" meant the perfecting of nature. John Slater also introduced the major debates over the kinds of literary art at stake in Lastanosa's work on history and antiquities, and Bruce Moran and Bill Eamon also spoke of the "art" that is involved in craftsmanship and what we would now call scientific practice. Perhaps, then, we need to begin our considerations by thinking of Lastanosa's "art" rather than his "science." If we do, it is possible to see almost all the papers as building a case for the following proposition: Lastanosa was not especially interested in what we today would call "science." Instead, he wished to work with nature in order to understand and "perfect" it. That is, in both verbal and material ways, he wanted to shape nature to moral ends rather than to explore its details for their own sake or for material advantage. Many of the purposes he found in nature were aimed a moral improvement, while others were aimed at utility and pleasure, but they appear to have been organised around "art" and "artfulness" in the older meaning of the terms. It fair to say, then, that many of the authors above have suggested that Lastanosa was interested in highlighting the things of nature (*res*) as evidence for the moral purposes inherent in God's creation.

In wishing to find evidence of moral teachings in the creation, and of perfecting our uses of nature in accordance with those teachings, Lastanosa shared his aims with many, probably most, collectors of *naturalia* in his time. But some other collectors, even in his generation, were beginning to organize their activities around a different set of purposes, separating moral history and philosophy from natural history and philosophy. It was those others, not those similar to Lastanosa, who are easier for us today to recognize as "scientists." In other words, if we consider the main theme of the conference, it would be true to say that Lastanosa combined art and science in his house and garden, in his religious life, and even in the items he collected; but this did not make him one of the innovators--one of the self-proclaimed neoterics or "new philosophers"--of his day.

It is then, in the relationship between the art and science of Lastanosa that we can begin to see his purposes and his similarities and differences with others of his time. tThis can be seen in one of his major activities: collecting and displaying natural objects (*naturalia*) and finely-crafted ones (*artificialia*). Over the last few decades, there have been many historical studies that have pointed to the importance of collecting for many kinds of investigations in the early modern period, including those we call scientific.[1] Because of their intimate association with collectors and museums, art historians have been in the vanguard of this movement, while historians of antiquities, books, gardens, literature, science, technology, and medicine have also taken an interest in collections.[2] Many paintings of early modern collections show works of art mixed with scientific instruments and *naturalia*, and this mixture of art and science has of course been noticed. The world of modern museums has been problematised by these studies, but the purposes of the recent world continues to frame our understanding. To put it another way, we often read back into early modern collections something like modern aestheticism, or art appreciation: for the collector, being surrounded by beautiful things made one's life not only more pleasurable but even better and more noble. This seems more or less straightforward when it comes to collecting art or antiquities, for as with excellent wine, fine food, or well-fashioned clothes, there is something in the world of taste that is clearly about "the finer things in life." It also, of course, has a social cachet: having discriminating taste is also a social marker, as Pierre Bourdieu made plain.[3] Valuing fine things is also often taken to indicate a discriminating moral judgment.[4] In other words, collecting art or antiquities seems to conjoin high social status, wealth, taste, and aesthetic value--and in this way Lastanosa was making a statement about his importance to Huesca as well as about his "scientific" interests.

When we look a bit closer at his activities, though, we can make a few distinctions. There are different kinds of knowledge embodied in different kinds of objects. Put another way, there are some differences between collecting to display one's taste and wealth and collecting to display one's command of information. The social status associated with ostentation and taste appears to have more to do with singular examples: the particular work of fine art or craftsmanship that impresses enormously because of the remarkable purposes and practices evident in it. This may also be true of remarkable singular instances from nature.[5] On the other hand, the discriminating knowledge that comes from the careful study of ordinary natural objects--such as measuring and mapping the distances between observable stars, carefully noting the ways in which hundreds of seashells or grasses are similar and different, teasing apart the intertwined tissues of animal bodies to make the organs visible to all--may have other aims. Or to put it another way, making an object of fine art is different than making a scientific discovery, and patrons and collectors understood that very well. To see Galileo as a courtier is of course quite right, but to see him as "producing" the moons of Jupiter in the same way that Bernini might have produced a sculpture is not quite right.[6] Is it really possible to make discoveries simply by the application of creative imagination and art to the world? Doesn't the world "bite back" in some way that constrains scientific discovery, that prevents one producing a new discovery just because one wants to?[7] Not all planets have moons, and the number of planets in our solar system is limited: otherwise we are simply "making something up." Nor can one make an object of antiquity without "counterfeiting" it. In short, when it comes to the content of knowledge and its relationship to kinds of truth, there are some important differences between art and science, between objects produced according to a patron's wishes—such as shaped boxwood in Lastanosa's garden—and objects brought to a patron's attention--such as local antiquities or exotic species he placed in it.

In other words, from what we can tell from the documents, Lastanosa's collecting appears to have been very purposeful when it comes to works of art and literature, but more haphazard with regard to things pertaining to *naturalia*, with the possible exception of his keen pursuit of alchemical experiments. This may well be because he did not cultivate the necessary expertise about *naturalia* either in his own studies or by patronizing knowledgeable scholars. One of the most important absences from his documented list of associates is medical advisers. Physicians and apothecaries were among the most important experts of the period when it came to the assemblage of cabinets of curiosity that contained large amounts of *naturalia*. After all, their livelihoods depended on their command of the information embodied in the products of nature they handled, which included foods and other preservatives of health, drugs and medicinals, and other useful and curious items found in nature. For instance, when in the 1550s one of the first known cabinets of curiosity was being assembled by Francesco I d' Medici, Grand Duke of Tuscany, many medical experts were present in Florence; when at about the same time, Hans Jacob Fugger of Augsburg also built a *wunderkammer* he employed the physician Samuel Quickelberg (or Quiccheberg) to help him. Medical people themselves even began their own collections. Fugger's collections inspired the famous Zurich physician and humanist Conrad Gesner to begin to assemble a cabinet of *naturalia*, while other famous collections were being put together by the physician and botanist Michele Mercati in Rome, the physician and botanist Ulisse Aldrovandi in Bologna, the apothecary Ferrante Imperato in Naples, and the apothecary Francesco Calceolari (or Calzolari) in Verona. In the low countries, the collections of Bernhard Paludanus, physician of Enkhuizen, became especially famous, many visitors declaring that because so many of the specimens came from the East Indies, his collection was superior to those in Italy. In Spain, too, the greatest collections of *naturalia* seem to have been associated with physicians such as Monardes, people who worked with physicians (among others) such as Benito Arias Montanus, or the King of Castile, who could draw on his experts from the Indies as well.[8] In other words, we need to bear in mind a point very well made by Carlo Ginsburg when discussing "clues": the expert adviser who had to determine whether a work of art had been done by a certain painter or not, like the expert on shells or butterflies, needed to have a keen and exacting eye for detail more than an appreciation of aesthetics.[9] But the documents do not show that Lastanosa cultivated such exacting expertise for *naturalia* himself, nor that he employed expert advisers to do it for him.

From various indications about the content of his collections, then, Lastanosa seems to have gathered together items that fit not so much with an interest in *naturalia* per se as with an interest in religion, the classics, and similar subjects associated with virtue, which we may assume were related to his humanist education. For example, the items mentioned in *Habitación de las Musas* include a few items from Asia that would have had great monetary value, presumably purchased from Portuguese sources (which until the early 1630s had dealings with Japan and still continued to trade within China). They include items such as the snails of mother of pearl with Chinese characters, the chest of Chinese work, a Japanese idol, and the ivory horn once belonging to the king of Japan or of India. But note the confusion in the description of the last: it is not quite clear where it comes from, nor whether the "king" of Japan was the emperor or the shogun, nor which of the many "kings" of India it might have been. The listing of accounts of information from Asia also mentions many authors of the sixteenth century, three maps from the early 17th century, and various books, also mainly from the 16th century with the addition of the 1628 Roman edition of Hernandes. Lastanosa therefore does not seem to have been particularly eager to find out about the nature of Asia.

If one compares these few and slightly out of dated items acquired by Lastanosa with those of his contemporary in Amsterdam, the elder Swammerdam (born one year before Lastanosa and dying three years before him), it is clear that Lastanosa's interest in *naturalia*, especially that from Asia, was far less keen. Jan Jacobsz. Swammerdam owned a flourishing apothecary's shop on Oude Schans at "The Star" (no. 18), close to the Montelbaenstoren and thus right by the harbor. On the upper floor of his shop he collected Chinese porcelain and everything else he could, including *naturalia*, which he purchased from sailors of the East-India ships and elsewhere. After his death in 1678, his sons and daughter argued about how best to dispose of the collection, for which they published a catalogue drawn up by his naturalist son Jan, which listed the objects on 142 pages of two closely printed columns, organized roughly into stones and minerals, plants, animals, and artificialia.[10] It included an amazing number of coins and medals; a strange, small Chinese image stamped in silver; a gold Japanese idol; a gold representation of Gustavus Adolphus; an artificial mouse with copper wheels and iron springs that could walk about; a Turkish almanac with colored letters; a Chinese almanac; silver from Mexico; bloodstones; pumice from Iceland; a sample of a "miraculous earth" called "fixed milk of the Virgin Mary"; three eagle-stones (said to be collected from eagle's nests, and of great medical efficacy); seventy corals; a branch of a tree called Rose of Jericho; two birds of paradise ("with feet," contradicting the common legend that they did not have any); edible swallows' nests; seven crabs from the Moluccas; a stone from Portugal that cured fevers (*lapis antifebrilis*); an Indian millipede; a sea star; some insects; a unicorn horn six feet, three inches long; and more and more.[11] Clearly the curiosity of both Lastanosa and the elder Swammerdam had similar roots. But the Dutch apothecary collected *naturalia* both much more extensively and much more exactingly than the aristocrat. He possessed, for instance, one thousand nine hundred shells.

If Lastanosa was not an expert on *naturalia* himself, he no doubt relied on agents to collect for him. It would be very interesting to know more about who they were, and where they traveled. Huesca is a long way from the main city port dealing with the colonies, Seville, or even from the important port city of Valencia. Presumably Lastanosa maintained good connections with Italy, where he certainly was in contact with Kircher--who of course had a keen interest in natural history.[12] The exhibition on Lastanosa mentioned that he had thirteen correspondents in eleven cities outside Spain. But can we compare his network with Nicolas Claude Fabri de Peiresc (1580-1637), from Aix-en-Provence--not so far away? Unlike Lastanosa, Peiresc cultivated the study of nature by becoming a painstaking investigator in his own right. And although he had more contacts in Catholic regions than elsewhere, Peiresc's network stretched all over Europe, with tens, even hundreds of correspondents.[13] Moreover, while Peiresc began with similar interests to those of Lastanosa in combining moral and natural philosophy through collecting and studying alchemy, by the 1630s, Peiresc was moving into new and dangerous philosophical territory by supporting the work of Pierre Gassendi on Epicureanism, for instance.

We can also compare Lastanosa to another Frenchman of his own generation known to have been interested in science, Henri Louis Habert de Montmor (c.1600-1679). Montmor, too, came from a wealthy family, in his case one related to the greatest families in France. Like Lastanosa, too, Montmor received an excellent classical education, and he became a great patron of scientists and philosophers, providing his clients with "an infinity" of machines and instruments with which he had stimulated his own curiosity in the teasing out of details in nature for thirty years. In the late 1650s, Samuel Sorbière prepared a plan for the organization of meetings in the form of nine articles, with the goals of the meetings to avoid "the vain exercise of the mind on useless subtleties; rather, one should always propose the clearest knowledge of the works of God and the advancement of the conveniences of life, in the arts and sciences that best serve to establish them." For such reasons, the Académie Montmor is sometimes called the forerunner of the Académie royale des sciences.[14] Whatever Lastanosa's achievements may be, then, he cannot compare with great French patrons such as Peiresc or Montmor (or with Dutch ones like Constantijn Huygens) for encouraging the new philosophy.

Finally, Lastanosa also seems to have avoided one of the implications becoming clear in the new philosophy of his day. The investigations of people involved with medicine were not only very important for focusing attention on the material details of nature, but sometimes for leading them to support ideas that bordered on philosophical materialism. For instance, in alchemy the idea that had been at its center--that the material substances supporting bodily life are expressions of deeper, immaterial forces which are interconnected by webs of sympathy and antipathy, and properly discerned through allegory--was being challenged by the middle of the 17th century, as materialistic alternatives being developed. Paracelsus and most of his followers, such as Oswald Croll, made much use of symbolism and explanations based on divine mysteries, not unlike the Baroque "typology" discussed by Canizares.[15] But even in the later 16th century, some writers were doing their best to strip spiritualist explanations away from the careful descriptive work of understanding chemical processes. In the work of the Dutch physician, technician, and philosopher Isaac Beeckman, for instance, one can see his appropriation of "atomism" to explain such things.[16] Many early 17th-century philosophers were deeply impressed by the power given by such concepts to explain chemical transformations without the need to invoke immaterial powers. The philosophical debates remained deep in controversy, but one of their consequences was to raise further skeptical questions about spiritualist explanations for chemical transformations.[17]

In other parts of the medical sciences, similar processes were at work. The case of William Harvey is an interesting example: he had travelled to

Catholic Padua to study, where the best medical investigations of his generation were going on, sometimes in an explicitly materialist context. When Harvey was awarded his doctorate in philosophy and medicine in April 1602 by count palatine, he swore and signed the required oath to the Pope's supremacy and all the articles of the council of Trent--although at his request this was left off the copy of the certificate with which he returned.[18] It is worth nothing here, too, that Harvey's main patron was Thomas Howard, 2nd Earl of Arundel, a scion of one of the greatest English Catholic families of the era, although Arundel remained nominally a Protestant after 1615. Maybe, then, a whiff of Rome hung about Harvey, which might even account for why Harvey was not appointed a royal physician-in-ordinary until December 1639, by which time the religious policies of Charles I and Archibishop Laud seemed to many to be bent on the restoration of the old religion. Harvey was also associated with the philosopher Robert Fludd--who was the first person to support Harvey's discovery of the circulation of the blood in print. Fludd also placed the heart at the center of all animal motion, including the passions. The place of the heart at the center of human experience was then preoccupying many religious thinkers, and would lead, in the middle of the century, to Catholic worship of the Sacred Heart. But Fludd was also a pantheistic materialist, and Harvey shows many of the same philosophical attributes, although they have usually been attributed to his sympathy for Aristotle. Another of Harvey's friends was the most famous materialist of the day, Thomas Hobbes, of whom Sir Geoffrey Keynes notes that his early treatise *De corpore* shares a theory of the senses with Harvey's notes to his work *De motu locali animalium*.[19] On his deathbed, Harvey left a ring to Hobbes in order to be remembered by him. Harvey's own philosophical attack on the idea of faculties and *spiritus* was one of the things that most bothered his opponents.[20] And finally we have the several reports that Harvey was a free-thinker who believed that suicide was a proper way out of personal suffering, perhaps even attempting suicide with laudanum first in 1652 and doing it successfully after a debilitating stroke in 1657. All in all, the more one probes, the more one can only agree with those biographers who have found Harvey's personal views to be a very elusive. But they certainly point to the power of philosophical materialism in his day.

In the work of Harvey, the anatomical work of Descartes,[21] and elsewhere, one also finds growing doubts about the connections between natural and moral philosophy. In the classical and Augustinian view, both are connected in the *logos*, the meaningful world created by God, where the knowledge of nature leads one to the understanding of the laws of creation, which are moral and purposeful as well as true. But to many of the new philosophers, such as Descartes and his followers, truths about objects and truths about moral purposes were grounded on different kinds of knowledge.[22] When Galileo famously said something similar in order to defend his views about Copernicanism, his enemies were furious. For people like Lastanosa, the Augustinian project of pursuing the one body of truth that governed the whole universe, both in its spirit and its embodiment, remained alive.[23] But as Egido notes in her paper, even Lastanosa's relatively small efforts in collecting *naturalia* were strongly criticized by Gracian as leading away from moral philosophy. For people of Lastanosa's time, then, there was a very real and serious danger arising from the study of natural bodies: the separation of moral knowledge from the study of nature. As to what Lastanosa might have thought about the disciplining of Galileo by the Roman Inquisition, or the controversies begun by the publications of Descartes, Harvey, Hobbes, and other philosophical *novatores*, we do not know. His silence is, however, notable.

To conclude, then, Lastanosa no doubt had the aims of most great patrons in mind when he collected *naturalia.* His interest displayed his virtue and knowledge and not only cultivated his own inner life but encouraged the study of the works of creation in others. He was, however, born in a generation that saw some very great changes in natural philosophy. There is no evidence that he took on board the arguments dividing natural philosophy from moral philosophy, much less its materialism, which was even beginning to change the kind of chemical speculations that seem to have interested him. Lastanosa's intentions seem rather have been--as they were on the part of so many people--to support the new science only in part: to put it to work for moral instruction, while avoiding its dangers. Perhaps he could do no more. During Lastanosa's years Aragon was in great difficulties: the expulsion of moriscos in 1610 meant that about 20 percent of the province's population evaporated, with consequently very heavy rents on the remaining farmers to support the landlords, and with consequently many defaults of landlords despite these new impositions, which meant lower incomes for middle class lenders and urban merchants, too. From about the time of the mid-century terrible epidemic of plague the kingdom experienced negative demographic growth of the remaining population as well as a bitter war with Catalonia, while from 1629 there was a depression of the Indies trade to Castile, with major knock-on consequences, particularly in very high inflation. Aragon did not recover from these blows until after Lastanosa's death. All this may very well have limited the range of Lasanosa's correspondence and his agents even more than the growing institutional conservativism of the church.

Lastanosa's example might, then, show just how difficult it was to import into northern Aragon information and ideas that were connected with more cosmopolitan commercial cities. Perhaps if he had come from Valencia, with its Mediterranean connections and important medical school, it would have been easier for him. At the same time, there is no indication that he wanted to become another Peiresc or Montmor, much less a Descartes or Hobbes. These were dangerous times for one's immortal soul, and Lastanosa--like so many others--seemed best content to do all he could to continue the kind of collecting that exemplified the values of Christian humanism.

Neither Lastanosa, nor Aragon or 17th-century "Spain," were intellectually "backward." But he does not seem to have been committed to supporting the kind of interest in nature--the analytical materialism--that was already recognized as something new and potentially dangerous. Lastanosa is a fascinating figure, and the persons he befriended, supported, and employed deserve further study as well. From seeing him as part of a larger world rather than simply an individual we will understand his world much better. But at the moment I would be surprised if further studies showed him to be moving toward the separation of moral and natural philosophy, which was happening in many discussions elsewhere: he seems to have aimed to keep them together as best he could. In this sense, at least, he stood on the side of classical learning rather than on the side of the study of nature for its own sake. Lastanosa subordinated knowledge to the purposes of art: the study and possible recreation of the moral ends of human nature. That humans are the product of material an inhuman nature alone he may perhaps never have thought possible.

Notes

1. For some important early examples, see G. R. De Beer, *Sir Hans Sloane and the British Museum* (Oxford: Oxford University Press, 1953); Silvio A. Bedini, "The Evolution of Science Museums," *Technology and Culture* 6 (1965): 1–29; Barbara J. Balsiger, *The Kunst- und Wunderkammer: A Catalogue Raisonee of Collecting in Germany, France, and England, 1565–1750*, dissertation, University of Pittsburg (1970); Hugh Honour, *The European Vision of America* (Cleveland: Cleveland Museum of Art, 1975); Arthur MacGregor, ed., *Tradescant's Rarities: Essays on the Foundation of the Ashmolean Museum 1683, with a Catalogue of the Surviving Early Collections* (Oxford: Clarendon Press, 1983); Oliver Impey and Arthur MacGregor, eds, *The Origins of Museums: The Cabinet of Curiosities in Sixteenth- and Seventeenth-Century Europe* (Oxford: Clarendon Press, 1985).

2. For some recent attempts to connect art and science, see Thomas DaCosta Kaufmann, *The Mastery of Nature: Aspects of Art, Science, and Humanism in the Renaissance* (Princeton: Princeton University Press, 1993); Horst Bredekamp, *The Lure of Antiquity and the Cult of the Machine: The Kunstkammer and the Evolution of Nature, Art and Technology*, German ed. 1993, trans. Allison Brown (Princeton: Markus Wiener Publishers, 1995); David Freedberg and Enrico Baldini, *The Paper Museum of Cassiano Dal Pozzo*, Series B--Natural History (London: Harvey Miller, 1997); Giuseppe Olmi, "Museums on Paper in Emilia-Romagna from the Sixteenth to the Nineteenth Centuries: From Aldrovandi to Count Sanvitale," *Archives of Natural History* 28 (2001): 157–78; Pamela H. Smith and Paula Findlen, eds, *Merchants and Marvels: Commerce, Science, and Art in Early Modern Europe* (New York and London: Routledge, 2002); Wolfgang Lefèvre, Jürgen Renn, and Urs Schoepflin, eds, *The Power of Images in Early Modern Science* (Basel: Birkhäuser, 2003); Pamela Smith, *The Body of the Artisan: Art and Experience in the Scientific Revolution* (Chicago: University of Chicago Press, 2004); Claudia Swan, *Art, Science, and Witchcraft in Early Modern Holland: Jacques de Gheyn II (1565–1629)* (Cambridge: Cambridge University Press, 2005).

3. Pierre Bourdieu, *Distinction: A Social Critique of the Judgment of Taste*, trans. Richard Nice (Cambridge, MA: Harvard University Press, 1984).

4. Harold J. Cook, *Matters of Exchange: Commerce, Medicine and Science in the Dutch Golden Age* (New Haven: Yale University Press, 2007), 1–41.

5. See esp. Lorraine Daston and Katharine Park, *Wonders and the Order of Nature 1150–1750* (New York: Zone Books, 1998)

6. Mario Biagioli, *Galileo Courtier: The Practice of Science in the Culture of Absolutism* (Chicago: University of Chicago Press, 1993).

7. I am aware of various problems that arise in using the word "discovery", as in using the word "science," but for the current purposes of distinguishing between making and finding, I think them suitable. For some considered thoughts about the "production" of scientific knowledge, see for example Ian Hacking, *Historical Ontology* (Cambridge, MA: Harvard University Press, 2002); Ofer Gal, *Meanest Foundations and Nobler Superstructures: Hooke, Newton and the "Compounding of the Celestiall Motions of the Planetts,"* Boston Studies in the Philosophy of Science (Dordrecht ; Boston: Kluwer Academic Publishers, 2002).

8. Cook, *Matters of Exchange*, 20–33.

9. Carlo Ginzburg, "Morelli, Freud, and Sherlock Holmes: Clues and Scientific Method," *History Workshop Journal*, no. 9 (1980): 5–36.

10. G. A. Lindeboom, ed & comp, *Het Cabinet Van Jan Swammerdam (1637–1680)* (Amsterdam: Rodopi, 1980).

11. Also see more generally on Dutch collections Ellinoor Bergvelt and Renée Kistemaker, chief eds, *De Wereld Binnen Handbereik: Nederlandse Kunst- en Rariteitenverzamelingen, 1585–1735* (Zwolle: Waanders Uitgevers/Amsterdams Historisch Museum, 1992).

12. Paula Findlen, *Possessing Nature: Museums, Collecting and Scientific Culture in Early Modern Italy* (Berkeley: University of California Press, 1994).

13. Peter N. Miller, *Peiresc's Europe: Learning and Virtue in the Seventeenth Century* (New Haven: Yale University Press, 2000).
14. Harcourt Brown, *Scientific Organizations in Seventeenth Century France (1620–1680)*, reprint, 1934 (New York: Russell and Russell, 1967); Wendy Perkins, "The Uses of Science: The Montmor Academy, Samuel Sorbière and Francis Bacon," *Seventeenth-Century French Studies*, no. 7 (1985): 155–62; David S. Lux, *Patronage and Royal Science in Seventeenth-Century France: The Academie de Physique in Caen* (Ithaca: Cornell University Press, 1989).
15. Bruce T. Moran, "Paracelsus, Religion, and Dissent: The Case of Philipp Homaguis and Georg Zimmermann," *Ambix* 43 (1996): 65–79; Bruce T. Moran, *Distilling Knowledge: Alchemy, Chemistry, and the Scientific Revolution* (Cambridge, MA: Harvard University Press, 2005); Jole Shackelford, *A Philosophical Path for Paracelsian Medicine: The Ideas, Intellectual Context, and Influence of Petrus Severinus: 1540/2–1602* (Copenhagen: Museum Tusculanum Press, University of Copenhagen, 2004).
16. H. H. Kubbinga, "Les Premières Théories «Moléculaires»: Isaac Beeckman (1620) et Sébastien Basson (1621): Le Concept d'«Individu Substantiel» et d'«Espèce Substantielle»," *Revue d'Histoire Des Sciences* 37 (1984): 215–33.
17. Harold J. Cook, "The Decline of Alchemy," *Alquimia: Ciencia y pensamiento a través de los libros* (Madrid and Seville, 2006), pp. 68-81.
18. Jonathan Woolfson, *Padua and the Tudors: English Students in Italy 1485–1603* (Cambridge: James Clark & Co., 1998), 21–23.
19. Geoffrey Keynes, *The Life of William Harvey* (Oxford: Clarendon Press, 1966).
20. James J. Bono, "Reform and the Languages of Renaissance Theoretical Medicine: Harvey Versus Fernel," *Journal of the History of Biology* 23 (1990): 341–87.
21. For more on Descartes' anatomical studies during the 1630s and 1640s, see Cook, *Matters of Exchange*, 232–37.
22. Theo Verbeek, *Descartes and the Dutch: Early Reactions to Cartesian Philosophy, 1637–1650* (Carbondale: Southern Illinois University Press, 1992).
23. See, for instance, Tomás Carreras y Artau and Joaquín Carreras y Artau, *Historia de la Filosofía Española: Filosopfía Cristiana de los Siglos XIII al XV* (Madrid: Real Academia de Ciencias Exactas, Fisicas y Naturales, 1939–43), 2: 101–75.

CHAPTER TWO

THE SEVENTEENTH-CENTURY SPANISH SCIENTIFIC CULTURE: SPAIN, AMERICA, AND THE STUDY OF NATURE[1]

ANTONIO BARRERA

This paper compares the work of Don Juan de Vicencio Lastanosa (1607-1684) and Alvaro Alonso Barba (ca. 1569-1662). Lastanosa was a landowner-collector, who lived in Huesca, Aragon; Barba was a priest-miner who lived near Potosí, Perú. They, in other words, lived in two distant regions of the Spanish empire. I wonder about these two people, did they share a common culture as a result of belonging to the same empire? Why a man in Huesca decided to make a collection of natural and man-made objects, and, eventually create a museum at his house? Why and how did he find those objects? Why a man in Potosí decided to gather information about mining activities and write one of the most important books on mining technology at the time?[2] Did they, both, share a similar culture of collecting, and similar practices for understanding and studying nature? Putting together Lastanosa and Barba is an arbitrary decision. There are very few links between them: they belonged to the same empire and their lives overlapped for some years; Lastanosa had a copy of Barba's book in his library; both collected curiosities; both published books in the 1640s.[3]

This paper compares Lastanosa and Barba and argues that the Spanish empire promoted, at least two alternative ways of understanding and studying the nature world. One of these ways is connected with scholars and humanists who were devoted to collecting curiosities, antiques, natural and artificial objects, and books as a way to bring closer to home the nobility of the first things. These scholars produced knowledge based on relationship between objects and such fields as mythology, religion, politics, history, and morality.

The other way was connected to traders, artisans, and entrepreneurs who sought to gather practical, and most times, empirical information about commodities. This people produced a specialized knowledge, for instance, knowledge about minerals, with the aim of exploiting natural resources and commodities. In this paper I suggest that this ways of understanding nature and of producing knowledge were related to cultural and social contexts.

The respective activities of scholars and entrepreneurs were not opposed: scholars, no doubt, also sought wealth and their collections helped them to project power and wealth; traders also collected curiosities, paintings, objects and books. Both understand that the production of knowledge was a collaborative activity.[4] There was, however, one key difference between the two groups: the way they defined knowledge. For scholars, like Lastanosa, knowledge of things resided in the past, in moments closer to the time of creation; these scholars sought universal knowledge and connections between signs, natural entities, and artificial things; these scholars looked to the past as a source of knowledge and models for life. Entrepreneurs, like Barba, sought practical and specific knowledge as, for instance, the study in detail of metals; entrepreneurs looked to the present and future. They made investments in businesses and expected gains in the future: for that they needed to understand the present. Barba studied present mining practices and sought to improve them. Lastanosa sought to study curiosities to reach back into the past and had a better knowledge of that primordial, pure knowledge of the beginning of the creation.

The Spanish Empire between 1607-1648

The years between 1607 and 1648 marked a time of transition for the Spanish Empire. The constellations of Spanish political dynamics changed rapidly in both Europe and America. In 1607, the year in which Lastanosa was born, the British arrived in the Chesapeake Bay and established Jamestown.[5] By 1617, Virginia was producing tobacco and, with its economic success, immigration increased: it averaged about over 8.000 people per decade from 1630s to the 1650s.[6] Between 1620s and 1630s, the British established also colonies in what would be called New England. The combined population of New England and Chesapeake Bay was about 58,000 people by 1660.[7]

The Caribbean was also an important destination of British immigrants, and with the establishment of the sugar and slave plantation system in the islands, British America had become an economic center in the Atlantic by the 1640s. The French established their presence in America in the 1610s; the Dutch established their presence in America in the 1620s. In the 1620s Spanish American kingdoms were also in a process of transformation: American Spaniards in Mexico and Peru were asserting their power within the administrative, judicial, and ecclesiastical structures of the colony against the Spanish Peninsulars.[8] In Europe, simultaneously with the European expansion in America, the Thirty Years War had begun in 1618 and ended in 1648.

During this period, Lastanosa came into his adult life. In 1626, Lastanosa married Doña Catalina and Gaston Guzman. In 1632 his mother, his grandfather, and his great-uncle died, which left him in full control of his inheritance and destiny. In the years1630s, as mentioned above, the English began to migrate in large numbers to Chesapeake Bay, New England, and the Caribbean; and France entered the war in the Thirty Years. By then, the British, but also the Dutch and French, had broken the hegemonic control of Spain in America and Europe during the first half of the seventeenth century.

In 1640s, the English Caribbean began to produce sugar and to traffic in slaves; New England connected its economy to the Caribbean sugar-slave economy, by feeding the population and selling them timber and other commodities. British America began a period of relative prosperity. In the 1640s, the American Spaniards in Mexico and Peru were sending less silver to Spain and investing that money in America. In 1640, Portugal seceded from Castile; and Catalonia tried to do the same; France invaded Aragon. In 1644, Lastanosa's wife for eighteen years, doña Catalina, died; the following year, Lastanosa published his *Museo de las medallas*. I argue that Lastanosa's high social position in Huesca and the historical events of the 1600s to 1640s shaped his collecting projects and his methods for collecting and studying curiosities.

Lastanosa's project, which consisted in his museum and his work in numismatics, was an attempt to reassert the local status of Huesca within the kingdom of Aragon, and Aragon's status within the Spanish monarchy. In the 1640, few knew, and he was not one of them, what would be the ramifications for the Spanish monarchy of the British, French, and Dutch expansion in America; of the Thirty Year's War; and of the increase power of the American Spaniards in Mexico and Peru.

Lastanosa's Project

In 1630, a pastor who was tending his flocks in the vicinity of the town of Tamarid found a large quantity of silver coins with Spanish characters. The news spread rapidly, and soon people began seeking coins in the region. People found many pieces of silver, and the findings "enriched many residents of Tamarid," Silver shops of Zaragoza melted many of these coins, and so the "inquiridores de la Antiguedad," as Lastanosa called the antiquarians, had few coins left to study.[9]

The discovery of Tamarid put two groups to compete for the coins buried in the region: the antiquarians and Tamarid's residents who saw an opportunity in the silver coins of becoming rich—by melting them. About this last group, we know little. They probably were pastors, like the one found the first coins, peasants, and artisans. This was a group of people who was probably able to appreciate the value of the coins as antiques, but preferred their value in silver. A significant number of people like the residents of Tamarid decided, throughout the sixteenth century, to seek their fortunes in America—the left Spain, the Old World, in search of gold, silver, and riches in the New World. I am not arguing that the actual residents of Tamarid left for the New World, probably few of them left because it was more difficult for people from Aragon to migrate to the New World than people from Castile. After all, American felt under the crown of Castile, and legally it was easier for Castilians to migrate to the New World. My point is simple. There are at least two cultures in place in Aragon, and the Spanish empire in general: the elite, high-culture of the "inquiridores de la Antiguedad," and the artisan, entrepreneurial culture of peasants and artisans. I further argue that they had a different relationship with time, the first group looked to the past, the second group looked to the future.

More is known about the inquiridores de la Antiguedad: Don Juan de Lastanosa Vicencio was one of them. He lived few hundred kilometers from Tamarid, in Huesca. His encounter with the coins of Tamarid would result in his life-long interest in coins, medals, books, and curiosities. Most antiquarians stayed in Spain: they project to collect antiques was a project intrinsically connected to the creation of a Spanish past that could be related to the 1640s situation of Spain.

The artisans and peasants who melted the coins and the antiquarians who preserved for their museums represent two different ways of understanding nature yet they shared a similar method—empirical method—to study it.

Don Juan de Vicencio Lastanosa and the emblematic world

In 1645, a year after the death of his wife, Dona Catalina Gaston Guzman (1612-1644), Lastanosa published his book *Museo de las medallas desconocidas Españolas*. The Discalced Carmelite Fr. Geronimo San Josef wrote the "Aprovación" of the book. Fr. Geronimo explained that the study of ancient things is important, first, because antique things were closer to the origin of things; second, because antiques could be considered the source and principle of the things that followed them; and finally, because antiques provided information about the past, and the past "represents that candid simplicity in which those in the past lived, whose memory, in opposition to the present malice, makes antiquity the best thing." Fr. Geronimo concluded by stating, that whatever the reason for the study Antiquity, "Antiquity has always been venerated; and for that reason, it is worthy of particular estimation those who seek to investigate, illustrate, in publish it."[10] This is exactly what Lastanosa did: he inquired, illustrated, and published about antiquity. He had begun in 1631, just when he finished writing his false genealogy.[11]

By the 1630s, Lastanosa had epistolary exchanges with the Duke of Orleans, who had a famous museum. Between 1637 and 1638, Lastanosa visited the Duke of Orleans in Paris; during the visit he purchased books and curiosities. By 1644, Lastanosa's museum was more than just a collection of art and natural objects. It was also a center for the study of the universe by making manifest the invisible relationships between natural, historical, religious, and artificial objects. Above all, Lastanosa's museum was a site that allowed he and his friends to collaborate in this enterprise of gathering objects and producing knowledge about them.

In the course of the years, a group of friends sent Lastanosa objects and information about them, or opened the doors of their museums and libraries for him. Some of these friends were the Count de Guimera, Dr. Bartolome de Morlantes, chief royal chaplain at the Church of Our Lady del Pilar, and who became an important scholar of classical traditions; Dr. Francisco Ximenez de Urrea, chronicler of the kingdom of Aragon, and himself a collector who would gather, according to Lastanosa, the largest collection of Aragon; the famous Francis Filhol in the town of Tolosa; Dr. Juan Francisco Andrés Uztarroz, with whom Lastanosa had a close relationship; Dr. Gaspar de Lastanosa; the *Licenciado* Juan Francisco Buena-Fe, *racionero* of the Church del Pilar; the Jesuit Jerome Garcia, a student, like Lastanosa, of Jaca coins; Dr. Juan Orencio de Lastanosa, brother of Vicencio and canon of the Church of Huesca, who sent regularly antiques to his brother Vicencio; Dr. Francisco Fuser, canon of the Cathedral of Barbastro; *Licenciado* Martin Miguel Navarro, who was secretary of the viceroy of Naples; Dr. John Joseph of Sada, *racionero* of the parish of San Lorenzo in Huesca; Don Ambrosio de San Juan y Latras; the Jesuit Balthasar Gracian, who sent antiques, coins, and curiosities to Lastanosa; Marquis Torres; Juan de Garriz, charge with paying the army of Aragon; and Don Josef Jacinto Clymente. Later Lastanosa expand his circle of friends to include people from Madrid, but his closest and more loyal circle of friends was from Aragon—friends from the nobility and clergy.[12]

This people had a particular quality: they were curious—at least, this is one of the qualities they ascribe to themselves. If they saw a figure at the root of a plant they were able to appreciate it as a marvel instead of using it "as the liars and old vain ladies because it [the root with a figure] promotes their deception."[13] Curious people were people who "yearns fervently for the public good and the benefit of the learned" by seeking and restoring antiques and saving them from ignorance, which is a "horrendous and abhorrent Monster."[14] Curious people, like Lastanosa and his friends, therefore were the people interested in understanding and preserving things from the past: antiques for they benefited the public good and the scholars.

Lastanosa's museum was one place to preserve antiques and study them; they provided his Aragonese circle access to knowledge, and this knowledge provide them with status and social power. The museum with its collection of antiques and books and curiosities conferred Huesca and, by extension, Aragon and Spain the special status of a site of knowledge and preservation of the past.

The Site of knowledge: the museum

The museum Lastanosa can be described in the same way that he described the museum of his friend Francisco Filhol: "cipher of curiosity, wonder of wonders."[15] (Lastanosa, Museum, 13). The term "cipher" means at the same time a kind of writing that uses a code and can only be understood knowing the key to the code; cipher also means an emblem, which represents symbolically something else. Describing Lastanosa's museum as a cipher of curiosities, then, it means that the museum was an emblem and a collection of signs whose key was in the hands of the antiquarians.

His museum, like those of so many others at this time, created a space where art and nature were connected with one another and reflected each other. Thus, in the museum Lastanosa there were several desks "pretended to be made of ebony and ivory;" there were two lizards made of bronze in "such a way that, if put next to real animals, they could be taken for real if they had been colored;" and "a bronze mouse so natural that it has deceived its enemies [cats], preying on it but its toughness has mocked them [the cats]."[16] The museum Lastanosa showed how art was capable of imitating nature and how art was able to decode the secrets of nature. Objects of art could thus pretend to be something else and deceived those who did not know the real nature of the object.

Similarly nature imitated art. For example, a fern "is wonderful this plant among curious people because by cutting its root through the female fern, it represents perfectly an imperial eagle with two heads and spreading wings."[17] In some cases, the viewer can not decide which is more important if the raw material of an object or the artistic work, as in the case of the jewels in which "one doubts" Uztarroz says, "what should be given the first place, the artistic work or the raw material [of the jewels]."[18] In this case, the artisan and the work of art were at the same level as nature and nature's works.

What makes the objects in Lastanosa's museum so wonderful was not the object itself, the painted desks, the plants, the jewels but the way they showed how art imitated nature and how nature imitated art. The museum Lastanosa, as did museums throughout Europe as the time, broke the Aristotelian division between art and nature. The Lastanosa museum displayed nature as an artist, capable of making its own creations, and the artist able to imitate nature, which means at the very least, that the artist was able to understand efficient causes of natural things. Knowledge of nature provided power to recreate and imitate nature.

Each object in the museum appeared in all its specificity and decontextualized; the objects, in the museum, were connected with other objects in a room that were equally unique and decontextualized. What was the criterion for Lastanosa to collect the objects of the museum? First, the objects were either beautiful or unusual or strange or older or all of the above. His collection had rocks, giant objects, strange objects, precious objects, objects of great artwork, objects with medicinal properties, objects with magical properties, ambiguous objects, objects that change, all of which made the pieces of his museum exceptional, strange and incredible.[19]

Second, the objects are arranged in such a way that they created ambiguities in how to classify them. For example, from the tip of a stone, an object in Lastanosa's museum,

> nace una mata de coralina, de grandor de una tercia de alto y un palmo de ancho, revestido todo el tronco y ramas de una corteza blanquísima, formando aquella desigualdad y hermosura que se representa en la hoja de la sabina; y es prodigio digno de reparo que esta planta haya podido nacer, crecer y alimentarse en una materia tan dura como es una piedra guija.[20]

Was this object a rock or a plant? What kind of plant was it that could grow on a rock? What kind of rock was it that could allow a plant to grow attached to it?

Art could achieve the same ambiguity: the museum had a desk that seemed to be made of ebony and ivory, but "the dexterity of brush, so ingeniously," had painted it to look like ebony and ivory. In the museum, there were also glass that some looked like crystal and others like agate. There were also mirrors reflecting lights that seemed "beautiful stars with the same beauty" that the stars show and are admired for, in the sky.[21] The play between art and nature, so skillfully displayed in Lastanosa's museum, help to focus the attention on the specificity of objects and their relationship with other objects: a bronze mouse, a real mouse, a cat, a mirror, reflecting lights, stars, ebony, ivory, and wood became objects of study in themselves and in their relationship with other objects.

The collection of Lastanosa, like that of his friends Francisco Ximenez de Urrea and the Count of Guimera, was a collection of beautiful, strange, interconnected, and ambiguous objects. The Lastanosa collection established—discovered, he would have said, relations between objects that emphasized the creative dimension of nature and the artistic skill of sculptors, painters, masons and carpenters. The type of knowledge that came from studying its objects of the museum was a universal knowledge based on invisible relationships between objects.

The purpose of the museum: a political project

The house was built from north to south, with the main facade toward the east. The sun faced squarely on the windows while passing, throughout the year, the signs of Libra, Sagittarius, Cancer, and then moved back to Pices.[22] These signs were associated with Venus, Jupiter and Neptune, which also explains the numerous references to Venus, Jupiter and Neptune in the Lastanosa museum. One of the balconies facing north had the initials "SPQR," which mean *Senatus Populusque Romanus*—The Senate and the People of Rome. In the garden, there was a pool that reflected Huesca and a Hercules holding a globe. This is one of the many connections the museum made between Rome, Huesca-Aragon, and Monarchy.

Those connections continued inside the museum. In one of the first rooms described by Uztarroz, he did described a box decorated with

> un escuadrón de infantería de figuras labradas excelentemente con armas y trajes diversos, que multiplicándose por la oposición de espejos, se aumentan sin número, y marchando en pasos concertado al son de clarín, imitado con una espineta, se van encontrando unos escuadrones con los otros, representando con toda perfección lo que se ve en un campo con ocasión de darse una batalla

and on top of the box there was a statue of Lucretia, "opening her chest with a dagger." On the sides of the box there were some shells, one of them was mother-of-pearl.[23]

The connection between Lucretia, the battle, and the shells was a reference to the purity of Lucrecia and the establishment of the Roman republic, which in the context Huesca-Aragon was an allusion to the Fueros of Aragon, granted in 1247 by James I of Aragon in Huesca—precisely. The theme of Lucretia was repeated in another room of the house.[24] (Uztarroz, "Overview," 92).

It seems to me that the Lastanosa museum was, then, a place for knowledge where humanists, nobles and the clergy, that is the political, cultural, and religious leaders of Aragon, came together to produce knowledge, and in doing so projected Huesca-Aragon as centers of knowledge and power. Lastanosa's own work illustrates the type of knowledge that emerged from the museum.

The coins and medals are a good reference for thinking about the Lastanosa collection. Medals and coins from the past are antiques, rare and hybrid objects. They have links with antiquity, mythology, morality, politics, and nature. Some medals, for example, had figures of horses, which, according to Lastanosa, represented Spain. Other medals had figures of crocodiles that represented Egypt. Similarly, coins had representations of gods, emperors, and kings that captured specific moments of history and mythology. Somehow all objects in the Lastanosa museum were like coins and medals: objects with two faces.

Some collectors like the Jesuit Kircher used his museum as a starting point to reflect on the wisdom of God and his actions on nature.[25] Other, like Lastanosa, used the museum—the knowledge produced by the museum—for political purposes, rather than religious purposes. Lastanosa's books on medals, *Medallas desconocidas* españolas, discusses Spanish medals, of course, but through them, he discusses the singularity of Spain, its antiquity, its political power, and the close connection between Spain and Rome. Similarly, in his *Tratado de la moneda jaquesa y de otras de oro y plata del reyno Aragon*, Lastanosa discusses coins from Aragon, and he begins with those made in Jaca, the old capital of Aragon, and thus he discusses, again, the singularity of Aragon, its antiquity, and its political power.[26]

Lastanosa studied the coins in their own singularity, as surely were the objects in the museum studied by others. He made drawings of the coins, and offered scales to determine their size (from the drawing); he studied archival records to determine their origins and values; he melted coins to determine the amount of metal in them. Lastanosa studied house each coin in its uniqueness. But such coin's singularity was only worth in a system of relationships—in relationship to politics, history, mythology, and so on. He somehow could not establish the complex system of relationships surrounding the coins and left that part of the enterprise to someone else.

He suggests, for instance, that the words of some of the coins he studied were written in a special alphabet.

> Las Notas, i caracteres de las Medallas Españolas, son tan estraños, i esquisitos, que aviendo hecho diligencias, no vulgares, para averiguarlos, por ver si podria formar un Alfabeto, i rastrear con el sus misteriosos secretos, me hallo impossibilitado de alcançarlos.[27]

Lastanosa tied to break the code of the language used in these Jaca coins. He compared these characters with characters in other languages. He found that some characters were similar to the characters contained in his work of the monk Otfrid Weissenburg (c. 800), "whose forms," Lastanosa comments, "draw" in their books Johannes Trithemius (1462-1516) and Giambattista della Porta (ca. 1535-1615), "pero como estas señales no tiene mas fuerça, ni significacion que el Antojo, i Arbitrio de quien las delinea, no pueden servir de regla, i documento para explicar las Medallas, porque no se entiende la significacion oculta, que contienen sus letras."[28] Lastanosa dismissed, then, this line of investigation as arbitrary. Neither were the Runic, Egyptian, Celtic, and Ethiopian alphabets useful to crack the code of the Jaca coins. That part of the project was to be developed by his colleagues, who could use Lastanosa research and see the coins in his museum.

Lastanosa, I argue, understood the objects of his museum in terms similar to his medals and coins. The curiosities stored at the museum had two faces, the visible and the invisible; at the same time, objects belonged to a large and invisible world of relations that connected those objects with mythology, art, religion, politics, history, and nature. Lastanosa's museum, and Lastanosa himself, belonged to an emblematic way of understanding the universe.[29] It is even possible to argue that the objects in Lastanosa's museum were like letters and as they form “words” with other object-letters, they produced moral, political, religious, mythological, natural, and/or artistic meanings.[30] Learning “to read” things and their relations promised to reveal an ancient knowledge, and therefore a more pure, noble, primitive, and secret knowledge that any specialized knowledge about objects. Ultimately, Lastanosa used his museum to cast an image of himself in Huesca, an image of Huesca in Aragon, and an image of Aragon in Spain.

Specialized knowledge—the knowledge of experts, by contrast, was imperfect knowledge because it was unable to capture the connections between things in the universe. In this regard, specialized knowledge did not constitute real knowledge because it was not able to make the connections between things and things and history, mythology, religion, politics, and nature. Thus, the knowledge of a miner, such as Alvaro Alonso Barba, was flawed and incomplete because only studied minerals,

Alvaro Alonso Barba and the specialized world of artisans

At the beginning of this paper, I suggest that the inhabitants of Tamarid, who found and melted silver coins, had something in common with those who chose to go to the New World and seek their fortune there. Barba mentioned, writing from America, "that the desire of returning to Spain, full of riches, is common to all who live in these parts."[31] Some people managed to go back to Spain—rich; many stayed and disappeared in America, as the son of Catalina Avila. Years later, when Catalina finally heard news about her son, she felt “such a great pleasure that she wanted to be like a bird and fly to see” him—sintió *tan grande alteración de placer que quisiere como pájaro volar a veros*.[32] Barba himself returned to Spain but decided to return again, already old, to America.

Barba was about thirty-seven years older than Don Juan de Vicencio Lastanosa; Barba was a priest who came to America before or around 1588; he became pastor of the parish of San Bernardo in the town of Potosi.[33] In 1640 Barba published his book, *El Arte de los metales*, a book that Lastanosa had in his library.

Barba was curious person, just as Lastanosa’s friends. Yet Barba defined a “curious person” in a different way than Lastanosa: for Barba, a curious person was not so much the antiquarian—someone who sought knowledge in the relationship of things but rather someone who sought knowledge in the singularity of things. A "curious miner" was a miner who, for instance, tasted the soil—its flavor—to determine the minerals present in its sample.[34]

For Barba, a curious miner would explore and visit sites looking for new things. Lastanosa and his friends visited museums and brought and sent objects to museums. Barba traveled the Andes looking for new things, for example, his curiosity led him to see a lake near the mines of San Cristobal de Achocolla, in the Lipez province, where "the water boils in the middle of the lake; there the water raises, sometimes just a little, other times a lot, with a big and frightening noise."[35] The water from this lake was transformed into salt; it was used for the benefit of metals; and some people had "experimented" with this water as medicine and found it a good medicine for dysentery.[36]

Curiosity for both Lastanosa and Barba moved people to learn new things and ask questions about what was unknown. Yet they called different people curios. In the case of Lastanosa, he does not use the word "curious' to describe the soldier who, in the castle of Monzon, found in 1644 a golden coin and, after searching around, he found more than "three hundred golden coins."[37]

Lastanosa called curious the antiquarian: the person who saved, preserved and studied the coins. Maybe because the interest of the soldier was practical, although Lastanosa did not say it, and was interested only in material wealth that the gold coin represented he was not a curious person. Barba, on the other hand, called the miner who tested the soil's flavor a curios person. The miner of Barba, and the soldier of Lastanosa had the same interest in finding gold. Yet the curiosity that came from studying American new-things, and the curiosity that came from studying Spanish antique-things were different. Knowledge for Barba did not start in antiquity, despite his continues references to classical authorities, but on the senses. The first chapters of his book are about colors, smells, and tastes of the land.

In his discussion of odors, for instance, he says that

> La variedad de olores, que entre las tierras se halla, no es menos de notar, que otras cosas que hacen admirable a la naturaleza. Huele bien la tierra ordinaria, quando despues de haver passado los calores del Estío, recibe el rocio de las primeras lluvias: cocio el calor en el tiempo seco la moderada humedad, que contenia la tierra (causa de que procede el olor bueno en todas las cosas que lo tienen) y mezclada con la primera agua, exhala, y evapora con el calor apacible, que todos experimentamos.[38]

Barba comments that "Apacible olor es el que echan de si las Minas de los Metales, que llaman Pacos, si otros medios Minerales no los acompañan, e inficionan: y este olor bueno no es pequeña señal de la riqueza que tiene sus piedras, o tierra que llaman Llampos,"[39] If anything, Barba's knowledge is based on his senses and the relationship he established between sensory qualities.

For Barba, in contrast to Lastanosa, curiosity leads to the systematic use of the senses to produce "greater knowledge."[40] For Lastanosa senses play an important role in his museum: mirrors, for example, showed stars reflected in a room and in another room, mirrirs projected objects in the air, but the sensory verification of those events in itself did not produce knowledge. Knowledge consisted in understanding the connections between the statue of Lucrecia and the images of soldiers reflected in mirrors "representing with perfection what is visible in the field of war:"[41] the creation of the Republic of Rome as a model for the kingdom of Aragon.

Despite the differences between Barba and Lastanosa, they shared an important practice in common: an empirical approach to the study of nature/art. Lastanosa melted his coins, drew them, weighted them, described them in their singularity; Barba smelled, tested, and looked carefully at the singularity of the soil. Both used an empirical method to understand objects. Why was this method use in one case to find invisible relationships among natural and artificial objects and in another case to find sensory relationship among natural objects is one of the important questions of the seventeenth century? I suggest, by looking at the cases of Lastanosa and Barba, that this empirical method in the entrepreneurial context of America became a method to study natural objects in their own singularity. This method in the humanist and Baroque context of seventeenth-century Spain became a method to study natural and artificial objects and their relationship with a primordial past.

Lastanosa and Barba represented two ways of understanding nature and knowledge: Lastanosa represents the search for a universal knowledge with political, moral, and religious purposes. It was a collaborative enterprise: Lastanosa gathered objects and studied coins for others to build on both his museum and his book. Barba represents the search for specialized knowledge—that of artisans, miners, pilots, and traders. He sought practical and specific knowledge and relied, also, on the activities of others. The search for specialized knowledge was also a collaborative enterprise. To the extent that Lastanosa work on coins was never concluded, his work, at the end, was closer to that of Barba than Lastanosa would have like to admit. So perhaps the research in museums was bound to end as specialize knowledge once it was clear the emblematic project of the scholars could not be accomplised.

Notes

1. I would like to express my gratitude to Mar Rey-Bueno and Miguel López-Pérez for organizing and inviting me to the international conference "Lastanosa, Art and Science in the Baroque," Huesca, Spain, May-June 2007. I would also like to thank the The Instituto de Estudios Altoaragoneses and the Provincial Delegation of Huesca for their support and the *inolvidables* days in beautiful Huesca. Finally, I thank also the participants for their comments and suggestions. This paper is based on my conference: I only added some footnotes. It should be read as a working paper.
2. Alvaro Alonso Barba, *Arte de los metales* (Madrid: Consejo Superior de Investigaciones Cientâificas, 1992). This book was first published in 1640 and was translated to English and German during the seventeenth century; it was translated to French during the eighteenth century. See Felipe Picatoste y Rodríguez, *Apuntes para una biblioteca científica española del siglo XVI, Colección Reprint* (Madrid: Ollero & Ramos, 1999), 22. Picatoste claims that the first edition is from 1600 but that seems to be a mistake.
3. Lastanosa published his books on medals in 1645, see Vincencio Juan de Lastanosa, *Museo de las medallas desconocidas españolas* (Madrid: J. R. Cayón, 1977). At that time he also began writing his *Tratado de la moneda iaquesa* but it was only published in 1681.
4. On collecting, see Lorraine Daston and Katherine Park, *Wonders and the Order of Nature, 1150-1750* (New York: Zone Books, 1998); O. R. Impey, and Arthur MacGregor, *The origins of museums: the cabinet of curiosities in sixteenth- and seventeenth-century Europe* (London: House of Stratus, 2001); and Susana Gómez López, "Natural Collections in the Spanish Renaissance," in *From private to public: natural collections and museums*, edited by Marco Beretta (Sagamore Beach, MA: Science History Publications/USA, 2005).
5. On Virginia, see Alan Taylor, *American Colonies* (New York: Viking, 2001), pp. 125ff; April Lee Hatfield, *Atlantic Virginia: intercolonial relations in the seventeenth century* (Philadelphia: PENN/University of Pennsylvania Press, 2004); and Karen Ordahl Kupperman, *The Jamestown project* (Cambridge, Mass.: Belknap Press of Harvard University Press, 2007).
6. James Horn, "Tobacco Colonies: The Shaping of English Society in the Seventeenth-Century Chasapeake," in *The Origins of Empire: British Overseas Enterprise to the Closa of the Seventeenth Century*, edited by Nicholas Canny, 170-192 (Oxford, New York: Oxford University Press, 1998), 177.
7. Taylor, *American Coloniees*, 365.
8. See Jonathan Irvine Israel, *Race, class, and politics in colonial Mexico, 1610-1670* (London: Oxford University Press, 1975), 83.
9. Vincencio Juan de Lastanosa, *Museo de las medallas desconocidas españolas* (Madrid: J. R. Cayón, 1977), 61-62.
10. Fray Gerónimo de San Josef, "Aprovación," in Lastanosa, *Medallas*, sin paginación.
11. See Manuel Alvar, "Una genealogía fantástica de los Lastanosa," in *Homenaje al profesor Juan Torres Fontes* (Murcia: Universidad de Murcia, 1987): I, 47-55.

12. For this list of names, see Lastanosa, *Museo*, 9. 10, 11, 12, 13, 59, 61, 62, 69. 70, 71, 72, 77, 78, 81, 102. 108, 109, 116.
13. Andrés de Uztarroz, "Descripción de las antiguedades y jardines de D. Vicencio Juan de Lastanosa," in *Don Vincencio Juan Lastanosa; apuntes bio-bibliográficos*, edited by Ricardo del Arco, 75-127 (Huesca,: L. Pérez, 1911), 116.
14. Lastanosa, "Museo," 54-55 and 31.
15. Lastanosa, *Museo*, 13.
16. Uztarruz, "Descripción," 80, 83, 84, 111.
17. Uztarruz, "Descripción," 116.
18. Uztarruz, "Descripción," 95.
19. Lorraine Daston and Katherine Park, *Wonders and the Order of Nature, 1150-1750* (New York: Zone Books, 1998), 273.
20. Uztarruz, "Descripción," 117.
21. Uztarruz, "Descripción," 82 and 100.
22. Uztarruz, "Descripción," 75-76.
23. Uztarruz, "Descripción," 79.
24. Uztarruz, "Descripción," 92.
25. Paula Findlen, *Possessing Nature: Museums, Collecting, and Scientific Culture in Early Modern Italy* (Berkeley, Los Angeles, London: University of California Press, 1994), 93.
26. See Lastanosa, *Museo*; Vincencio Juan de Lastanosa, *Tratado de la moneda iaquesa y de otras de oro y plata del reyno de Aragón* (Madrid: Juan R. Cayón, 1977).
27. Lastanosa, *Museo*, "A los lectores," sin paginación.
28. Lastanosa, *Museo*, "A los lectores," sin paginación.
29. William B. Ashworth, "Natural history and the emblematic world view," in *Reappraisals of the Scientific Revolution*, edited by David C. Lidberg and Robert S. Westman (Cambridge: Cambridge University Press, 1991).
30. On this idea, see Jan C. Westerhoff, "A World of Signs: Baroque Pansemioticism, the Polyhistor and the Early Modern Wunderkammer." *Journal of the History of Ideas* 62, no. 4 (2001): 633-650, 634.
31. Alvaro Alonso Barba, *Arte de los metales* (Madrid: Consejo Superior de Investigaciones Cientâificas, 1992), 30.
32. Rocío Sánchez-Rubio e Isabel Testón-Núñez, ed. *El hilo que une: Las relaciones epistolares en el Viejo y el Nuevo Mundo (Siglos XVI-XVIII)* (Mérida: Universidad de Extremadura, 1999), 45.
33. José M. López Piñero, Thomas F. Glick, Víctor Navarro Brotóns, and Eugenio Portela Marco, *Diccionario histórico de la ciencia moderna en España*, 2 vols, (Barcelona: Ediciones Península, 1983), I, "Barba, Alvaro Alonso."
34. Barba, *Arte*, 9.
35. Barba, *Arte*, 15.
36. Barba, *Arte*, 15.
37. Lastanosa, *Museo*, 114.
38. Barba, *Arte*, p. 6-7.
39. Barba, *Arte*, 7.
40. Barba, *Arte*, 9.
41. Uztarruz, "Descripción," 79.

CHAPTER THREE

FROM *HISTORIA NATURALIS* TO HISTORIA *AU NATURALE*: LASTANOSA AND THE NAKED TRUTH

JOHN SLATER

Pliny does not mince words when describing historiography to Vespasian in the dedicatory to the *Historia naturalis*; it is hard work to make old things seem new, to give novelty the weight of authority, to give brilliance to the obsolete, and illuminate what time has darkened:

> res ardua vetustis novitatem dare, novis auctoritatem, obsoletis nitorem, obscuris lucem, fastiditis gratiam, dubiis fidem, omnibus vero naturam et naturae suae omnia. itaque etiam non assecutis voluisse abunde pulchrum atque magnificum est.[1]

Two elements of this description—the difficulty of the task and the ultimate goal of illumination—become commonplaces among early modern Spanish historians. Juan Cortés Osorio cites this very passage in praise of José Vicente del Olmo's *Relacion historica del auto general de fe*, saying that Olmo had overcome "las dificultades que uno de los mayores Sabios de Roma propuso, si no como invencibles, como arduas: *Ardua res est* (dize Plinio) *vetustis novitatem dare*. . ."[2] Juan Francisco Andrés de Uztarroz echoes it in the dedicatory to his *Vida de San Orencio*.[3] For his part, Vincencio Juan de Lastanosa takes the entire passage as the epigraph to his work on numismatics, *Museo de las medallas desconocidas españolas*.[4]

Given its elegance and concision, the popularity of Pliny's passage is not surprising. But Lastanosa's use of it as an epigraph is worth considering for two reasons. The first is that Lastanosa forthrightly declares that he has very little light to shed on the obscure meanings of the coins, explaining that despite his considerable efforts to read their characters, "me hallo imposibilitado de alcançarlos."[5] In other words, there is no illumination; he publishes the images of the coins, but does not himself use them to reconstruct a narrative of the past. He calls his interpretative failure the "obscuridad del Asumpto" a phrase he uses in his dedicatory to Bernardino Fernández de Velasco and again in the "Advertencias al Museo."[6] This is not an extravagant example of false modesty; the aim of Lastanosa's book is to encourage *other* scholars to take up "las plumas en honra de nuestra Nacion, escribiendo varias ilustraciones."[7] Where Pliny describes his own struggle to bring light to the obscurities of the past, Lastanosa only lays the groundwork for future illumination.

The second reason that Pliny's description of the hard work of writing history makes a somewhat peculiar epigraph to the *Museo* is that Lastanosa makes a telling error in translation:

> Cosa es dificultosa dar novedad a las cosas viejas, autoridad a las nuevas, lustre a las desnudas, resplandor a las que están en tinieblas, gracia a las infelizes, i credito a las dudosas; pero mas dificil es dar su Naturaleza a todas las cosas, y dar a la misma todas sus propiedades secretas. I assi quando esto no se pueda alcançar, o hazer; solo aver querido, es obra grandemente magnifica.

Even a cursory comparison makes it clear that Lastanosa based this passage on Jerónimo Huerta's Spanish translation of the *Historia Naturalis*, which he owned, although he made a few alterations and additions.[8] Among these changes is Lastanosa's phrase "resplandor a las que están en tinieblas" for Huerta's "luz a las escuras" (Huerta 3). The former is somewhat more grandiose, if no more accurate; it shows Lastanosa's desire to elevate the style of the translation. The most striking thing about Lastanosa's version, however, is the phrase, "lustre a las desnudas." Huerta translates Pliny's original—"obsoletis nitorem"—quite reasonably as, "lustre a las desusadas." The substitution of the word "desnudas" for "desusadas" makes no sense; it must either be an error in translation or typography. It would seem that in Lastanosa's attempt to elevate the style of the passage, the meaning of "obsoletis" itself gets lost.

There is reason to believe, however, that something more than a typographical error is at stake. Like the "obscuridad del Asumpto"—the indecipherability of the coins' characters—the use of "desnudas" both suggests problems of translation and points toward broader issues in Baroque historiography. References to nudity in a variety of contexts are abundant in the period's historiographic manuals and debates: in allegories of Clio, discussions of the "naked truth," and references to an austere, uncomplicated style. In other words, the question of nudity, both in terms of historical substance (how to tell the naked truth) and style (whether history should be adorned with eloquence or stripped of excess) was of vital interest to seventeenth-century historians.[9] He may not have been a historiographic theorist himself, but Lastanosa's friends—from Juan Francisco Andrés de Uztarroz and Jerónimo de San José to Francisco Fabro Bremundan and even Francisco de la Torre y Sevil—had a great deal to say about how truth of a historical narrative might be manifested through style, whether naked or adorned.

If the use of the word "desnudas" can be taken as a subtle nod to the undercurrent of historiographic debate in Lastanosa's circle, the structure of the *Museo* demonstrates a more direct engagement with contemporaneous historiography. The book can be divided into four parts. First, there are about forty-five pages of front matter, including Lastanosa's "Prefacion" in which he lays out goals and offers his conviction that the coins are autochthonous (his only substantive historical claim). This is followed by a second section: fifty-five pages printed only on the recto, consisting of engravings of the coins, totaling twenty-three pages of plates. In the third, Lastanosa adds approximately sixty pages of descriptions of the coins. Lastly, three authors—Pablo Albiano de Rajas, Juan Francisco Andrés de Uztarroz, and Francisco Ximénez de Urrea—write preliminary "ilustraciones." So, of the roughly 225 pages of text, Lastanosa contributes approximately one hundred pages generally related to the coins' interpretation, almost as much as the other three co-authors combined. Each of these sections, as we shall see, respond somewhat differently to the various tensions within the evolving field of Baroque historiography.

While the stylistic and structural characteristics of the *Museo* have not generally led scholars to consider the work as precisely historical, Lastanosa's book *can* be seen as a kind of experimental history. As a work of history that lacks a historical account, the *Museo* permits an exploration of what happens when one of the necessary elements for the creation of history, a narrative of events, is removed. Lastanosa's own writings about the coins and his careful division of the book into four discrete sections, when seen in light of his curious translation of "obsoletis nitorem," suggest that *Museo* actively negotiates the stylistic debates within contemporaneous historiography. To explore how it does so, I will look first at the nature and tenor of the historiographic debates within Lastanosa's circle and throughout Spain, before returning to the *Museo* to show that "desnudas" is best seen as a revealing decision, rather than a careless error. In fact, in its responses to a variety of conflicting demands, the *Museo* itself lays bare a number of conventions that are increasingly important to the work of creating narratives about the past during the course of the seventeenth century.

I. Clio Clothed and Exposed

Lastanosa's library contained the two most important historiographic manuals of the seventeenth century: Luis Cabrera de Córdoba's *De historia, para entenderla y escribirla*, and Jerónimo de San José's *Genio de la historia.*[10] The former contains an allegory of Historia borrowed from Cesare Ripa's *Iconologia*, to which Cabrera makes a telling change.[11] While Ripa's figure is clothed, Cabrera's is "desnuda, porque la simplicidad le es natural."[12] Nearly every historian of the seventeenth century likely would have agreed that history ought to tell the naked truth, even if they lamented that this was infrequently the case in practice.[13] Gonzalo de Céspedes y Meneses, for example, whose *Historia apologetica* Lastanosa owned, commands historians to write "verdades libres, *desnudos* los buenos, o los malos aciertos."[14] Francisco Pinel y Monroy even suggests that the most naked truth—"la verdad más desnuda"—is to be found in the raw materials of history, documents and manuscripts, rather than in historical narrations.[15] But references to nudity in the "metatexto historiográfico" have a variety of functions beyond those found in allegories of Clio or discussions of the naked truth.[16] As Cabrera's invocation of simplicity indicates, recourse to tropes of nudity among historians is frequently a means to discuss style, rather than substance.

The careful distinction between historigraphic substance and style, matter and form, is commonplace during the seventeenth century. Antonio de Solis y Rivadeneyra explains in his *Historia de la conquista de Mexico* that "Los Adornos de la Eloquencia son accidentes en la Historia, cuya substancia es la Verdad."[17] Cortés Osorio similarly distinguishes between "la substancia, y el modo de [un] libro."[18] This kind of metaphysical language may seem a long way from stylistic nudity, but in the *Constancia de la Fee* Cortés Osorio likens substance and mode, on the one hand to a body and clothing, on the other.[19] From this analogy, it is a small conceptual step to understand the substance of history as either decorously dressed or stripped of ornament.[20]

There were a number of ways, beyond the allegorical, that historical style might be characterized as naked. The simplest of these was the absence of notes and references, that is, the literal blankness of the margins. San José, a close friend of Lastanosa's, explains that his *Genio de la historia*, contains "escritura desnuda," which is "libre del embarazo de las margenes."[21] Manuel Mariano Ribera similarly praises Catherine of Sienna's *Ramillete de epistolas y oraciones* (1698) for its "verdades desnudas de ornatos criticos."[22] It was more common, however, to oppose nudity and simplicity to adornment and sublimity in the style of the narration itself. Alfonso Sánchez, for example, flatly states that history must be "brevem, nudam, veram," that is brief, naked or unadorned, and true.[23] He explains his aversion to embellishment by saying, "omnem exquisitum, & curiosum ornatum detraho" [I strip away all finery and elaborate ornamentation] and dismisses digressions as "fucus" or rouge; nothing more than a cosmetic that covers the truth. Francisco Fabro Bremundan, another friend of Lastanosa's, lays out the central issue for historians who favored a sparser style, stripped of rhetorical flourish; in his dedicatory to María Luisa de Austria, he remarks, "bastava a las grandes Hazañas su mesma desnuda verdad, sin necesitar ellas del adorno de expressiones exquisitas, y mendigadas del Arte."[24] In other words, for Fabro Bremundan, any affectation on the part of the historian simply obscures the greatness of the events described.

The opposing view, that the achievements worthy of history should be clothed in the sublime style, is evident when comparing Fabro Bremundan's "grandes Hazañas" with Cabrera's "heroycas hazañas" in the following passage: praising Cristóbal Suárez de Figueroa's *Historia y anal relacion*, Cabrera says, "solo de su estylo se podia, y devia fiar la inmortalidad de tan heroycas hazañas."[25] The subtext is that history should be clothed in a language that approximates—or intimates through style itself—the magnitude of the events narrated. He is even clearer on this point when comparing history and painting in the *De historia*: history, like painting, is obliged to reveal the faults of the wicked—"desnuda las personas viles"—but must also cover "las nobles con propiedad de vestidos."[26]

Cabrera pertains to an earlier generation and is decidedly not a member of Lastanosa's circle.[27] However, this does not mean that there were not proponents of clothed history among Lastanosa's favored writers and friends. Juan Briz Martínez, for instance, whose *Historia de la fundacion, y antiguedades de San Iuan de la Peña* is cited in Lastanosa's *Museo*, condemns those who "se contentan, con contar *desnudamente* los sucessos, sin saber engrandezer, y dar lustre a las cosas."[28] Instead, embellishment should be added to a historical narrative "como lo haze el ingenioso Bordador, que adorna el vestido, sin menoscavo de su figura, con maravillosos recamados, de oro, perlas y margaritas." Another of Lastanosa's friends, Francisco de la Torre y Sevil, similarly balked at the idea of an austere narration.[29] In his *Luzes de la Aurora, dias del sol,* Torre y Sevil defends himself from those who find his style "estrañamente florido para historia" by saying that it could have been dressed worse: "Bien pudiera vestir, aunque desaliñadamente, toda la narracion de equivocos."[30] In short, the proponents of a more decorous Clio, tastefully clothed in rhetorical adornment, believe that the historian should use language commensurate with the greatness of the events described. A spare style simply cannot do justice to heroism.

Given the prevalence and intensity of the debate over nudity, it is difficult to imagine that Lastanosa's peculiar translation of Pliny would not have been read in this context. In his library and among his friends there exists a sartorial shorthand to talk about issues that went to the heart of what history ought to do and what it could be. On the one hand, Fabro Bremundan, Sánchez and others urge for restraint, for narrative nudity. On the other, Cabrera, Briz Martínez, and Torre y Sevil argue for historiographic *habillement.* This debate is not simply about aesthetics and has almost nothing to do with sex. As we shall see, the question of nudity has much more to do with the role of the historian and his relationship both to his readers and his subject than any erotic musings about Clio.

II. Modesty, Propriety, and Readership

To this point, I have largely discussed the way images of nudity and clothing are employed in the "metatexto historiográfico," but looked only briefly at the practical implications of this figurative idiom. In fact, these images form part of a much larger debate over historiographic style, one that can be explained in terms of two oppositions: brevity versus abundance, and the sublime versus the laconic style. When historians speak of favoring austerity or nudity, they align themselves with the proponents of brevity and with the incisive, laconic style frequently associated with Tacitus and Juan de Mariana. Those who favored an adorned, clothed style implied their preference for the sublimity and expatiation generally attributed to Tully and the sixteenth-century historian Jerónimo Zurita. However, the terms "sublime" or "laconic" were often used differently. For some, "laconismo" referred only to the absence of digressions and amplification; for others it indicated linguistic simplicity. An exact taxonomy of stylistic schools is difficult because there was overlap among them. Nearly all historians of the period express approval of Tacitus, for example.

It is in this context that images of nudity and adornment are again helpful because they suggest a related set of terms that describe the historian's own role within the narration: modesty and propriety. These terms operate in ways somewhat counter to what one might expect. The proponents of naked history believed that this historian should himself be modest. San José, for example, shows discomfort with the idea that the authority of history might reside wholly in the person of the historian. It is the verisimilitude—a word that appears several times in the *Museo*—of the narration that should convince the reader through reason. Cabrera's notion of the historian's task is different; for him, history must move its readership through an appeal to the emotions. As I mention above, Cabrera is much more concerned with the propriety of the narration, its stylistic commensurateness with events described or "decorum" in literary terms.

These are the choices that faced Lastanosa as he writes the *Museo*. On the one hand, naked history that foregrounds the plausibility of the narrative; through the use of a simpler style the historian modestly refuses to draw attention to himself. On the other, clothed history that underscores the authority of the historian; readers are overwhelmed by demonstrations of rhetorical mastery that themselves signal the importance of the events. Within Lastanosa's circle, Uztarroz favored an eloquent "estilo copioso," while San José's *Genio de la historia* encouraged the avoidance of sublime style.[31] These variances often amounted to differences of emphasis, rather than ideological opposition.[32] In fact, both Uztarroz and San José argue on roughly the same grounds. Uztarroz claims in the *Progressos de la historia* that too often, "es tanta la concission [sic], que no se comprehenden los sucessos, aunque por mayor se saben."[33] Taking an opposing tack, San José strongly cautions against the use of flowery language, which "en muchas horas el mas atento no la pueda entender."[34] Evident in both Uztarroz and San José's complaints is a concern with the reader's understanding; the wrong style can lead to incomprehension and derail history's moral instruction.

To ensure that readers properly understand the magnitude of historical achievements, partisans of propriety rely on digression, amplification, description, and what Cabrera calls "variedad, en el escoger las cosas, y las palabras, y en su buena disposicion y asiento."[35] For Cabrera, there is no *stylistic* difference between history and poetry. He says that history should be narrated "con ornamento y gala, como le toca, y no de otra manera," and flatly states that "[e]ngañanse los que piensan ser historia sin artificio."[36] Ornament and embellishment serve two purposes: first they help the historian persuade readers more forcefully through the use of visual images; and second, they give "claridad" to the narration, by creating a richer, fuller backdrop for the events of history.[37] It is perhaps not surprising that it was precisely his overwrought, even "gongorista," style that drew the censure of his critics when Cabrera published his biography of Philip II.[38]

Despite these criticisms, Cabrera's style had many adherents. One of these is Antonio González de Rosende, whose *Vida i virtudes del ... Señor Don Iuan de Palafox i Mendoza* is at times wildly ornate. In a "censura" to the biography, Juan Zafrilla y Azagra explains that history has the obligation to "adornar" the virtues of Palafox with "varios, i hermosos discursos," and to "inclinar los afectos para la imitacion de vida tan inculpable, i tan ejemplar."[39] It is precisely this aspect, the desire to move the reader to imitation, that Uztarroz mentions in a *censura* to Baltasar Gracián's *El Criticón*. He says that heroic acts, when related in "cláusulas breves, tal vez se confunden y quedan defraudados los hechos dignos de la memoria."[40] Brevity both robs the events their due and cheats readers of the instruction they seek. Stylistic shortcomings amount to injustice, a lack of propriety.[41]

But propriety is not only a question of properly calibrating style and subject. It is also a question of being mindful of history's readership: princes. As Uztarroz's friend Tomás Tamayo de Vargas explains: "La Historia se institutó para enseñaça de los Principes, porque encendidos sus animos con la luz de las hazañas de sus predecessores o iguales, se adelanten en su imitacion."[42] Thus, history should be the narration of great events, told in an elevated style, to an elite few who are capable of understanding the lessons of the past and who possess the power to act on them. In this sense, the propriety of history grows out of its mediating role between the great men of the past and those of the present.[43]

San José has little patience for this. He finds authors who write in the sublime style to be fundamentally narcissistic, "usurpando, con la gala del decir, la atencion a las cosas, i robando los aplausos, i admiracion, que se debia principalmente a los sucesos referidos."[44] By writing in the sublime style, complains San José, the historian, and not the history, becomes the object of the reader's attention. His more modest approach is no less ambitious, but posits a very different audience: a reader of "mediano ingenio."[45] In other words, the stylistic register should be determined not by the nature of historical events but by the abilities of an average reader.

In order to reach the reader of "mediano ingenio," San José counsels the use of an "estilo medio" which is "ni mui alto, ni mui baxo; sino con una grave, lisa, i corriente mediana."[46] (162). Once again, commensurateness is the key, but it is the reader and not the events that should guide the historian. San José's characterization of the excesses of the sublime style is damning: "Una metafora sobre otra metafora, i en cada palabra diez figuras, i en cada figura quince alegorias, i alusiones, que el mismo a quien esta obscuridad afectada costó mucho estudio, i desvelo, despues de escrito no lo entiende, ni sabe lo que quiso decir."[47] (147-48). He even inverts Pliny's "obscuris lucem," explaining that what the sublime style actually does to the Spanish language is "obscurecer su lucimiento."[48] Where Cabrera attempts to move the reader, San José tends to focus on the ability of narrative verisimilitude to establish authority. Because it is the plausible sequence of events that constitutes the source of a history's perceived reliability, interruptions in the form of long digressions do little to make history more credible.[49] A crude summing up of the differences between Cabrera and San José would be that Cabrera makes an emotional appeal through the elevation of language, while San José makes a reasonable appeal through the verisimilitude of narration.[50]

San José and Uztarroz remain friends, although they have meaningfully different stylistic agendas, as the difference between Uztarroz's "estilo copioso" and San José's appeal to the "mediano ingenio" shows. Affinities for a particular school of historiographic style are generally expressed through a partiality for a particular Roman historian—Tully over Tacitus, for example—or by praising a historian from a preceding generation, as Uztarroz shows in his preference for Zurita over Mariana. For all historians, however, the question of style was inescapably related, on the one hand, to the more complete and efficient communication of truth, and greater didactic efficacy, on the other. This was the case whether history was adorned or naked.

III. History and the *Museo*

While Lastanosa's friendships with Uztarroz and San José place him at the center of the social circle that discussed issues of historiographic style, his own writings are difficult to relate to this context. With no real narrative, Lastanosa's efforts seem to correspond more closely to the broad conception of "Historia" described by Gianna Pomata and Nancy G. Siraisi than to the narrower definitions of history seen thus far.[51] However, as the epigraph from Pliny demonstrates, Lastanosa is not without historical aspirations.[52] As I have said, the *Museo* can be seen as a historiographic experiment. After all, Lastanosa has a great deal of documentary evidence in the coins themselves. He has a clearly defined goal and audience—the scholars who will one day write the history he cannot—and he understands very well the rhetorical issues at stake. If we grant Pinel y Monroy's contention that the most naked truth lies in raw materials and not narration, we can even say that Lastanosa has access to "la verdad mas desnuda."[53] The only thing missing is the ability to make sense of the coins.

This limitation, when combined with the "desnudas" of Lastanosa's epigraph, would seem to make modesty and nakedness the likely choice for the *Museo.* And in many ways, it corresponds to this school. Lastanosa offers up the images of the coins "anelando ardentissimamente por la utilidad publica" and in the hopes that "los Inquiridores de la Antiguedad las ilustren, i divulgen."[54] Desiring, as he says, only the public good and leaving the interpretation of the coins largely to others, he seems to fit the conception of the modest historian San José is to outline when he publishes the *Genio de la historia* some years later.[55] Lastanosa's major concern is that the coins are neither surrendered to the "manos de la Ignorancia, horrendo, i detestable Monstro," nor returned to the "tinieblas del olvido."[56]

In reading the book, and attempting to compare the images of the coins to their textual descriptions, another of Lastanosa's choices becomes apparent. He does not juxtapose the images to his own text or imitate emblematic literature. Instead, his descriptions are retired to a latter portion of the book and grouped together.[57] It is actually rather cumbersome to flip back and forth between the section containing the text and that of the images, which tends to force readers to consider the plates in isolation. The effect is to reproduce physically the remove of the modest historian. The coins are allowed to stand on their own, naked, without the mediation of the author's voice.

The engravings themselves are rather stylized; like botanical illustrations they tend to represent ideal types rather than particular coins. Lastanosa defends this, explaining that readers will find "algunas Monedas mejoradas" but only because the originals are so hard to make out.[58] He does one further thing to facilitate the "inteligencia, distincion, i claridad a las Medallas," which is to add "algunas Notas, i Advertencias," or, descriptions of the coins. This is "necessario, i aun forçoso" if future historians are to overcome the "los estorvos, i embaraços" of deciphering.[59] Such a claim is hard to square with the descriptions themselves. Not only because these are occasionally erroneous, but also because they sometimes seem like pointless opportunities for digression.[60] For example, Lastanosa occasionally cites the early modern poet Luis de Góngora, as Karl Ludwig Selig contends, "to prove or refute the traditional description of the medal."[61] In practice, however, Góngora's poems themselves do little to help the reader's understanding. They are at times so difficult that they demand significant explanation—as in the description of the second coin—and at others so brief that the original meaning of the poem is lost, as is the case with coin 117.[62] As Aurora Egido explains, Lastanosa is less interested in using these poems to describe the coins than he is in creating a specifically Spanish cultural context in which to understand them.[63] He expresses the conviction, throughout the book, that the coins "han de ser proprias de España, i no de otra nacion," and it is in this light that the title—*Museo de las medallas desconocidas españolas*, should be understood.[64] This confirms the historiographic nature of Lastanosa's goal and his desire that they be used to formulate a history of Spain. But it also means that the descriptions have less to do with the appearance of any particular coin—the engravings, after all, are not particularly accurate—than with the author's attempt to dictate the terms for future interpretation. Whatever the coins are ultimately determined to mean, they must, implies Lastanosa, be seen as Spanish.[65]

Thus, both Lastanosa's limitations and desires put him front and center in the "Advertencias." In the absence of a historical narrative, the author's own convictions take the place of a plausible account. His inability to read the coins also limits his audience to those who can: the future scholars who will dedicate their "plumas en honra de nuestra Nacion." By employing sublime poetry in the service of his nationalist program, Lastanosa demonstrates a deep interest in determining the propriety of future interpretative outcomes. The coins, he is certain, tell of a glorious past just as the poems testify to the greatness of his own age.[66] This puts the descriptions, if not the entire *Museo*, in Cabrera's camp; while Cabrera is condemned for his "gongorista" style, Lastanosa goes straight to the source, making Góngora an important feature of the work.

Recalling Gracián's allegory of Historia—half ancient and wrinkled, half fresh faced and young—the *Museo* exhibits the influence of both naked and clothed styles of history without fully declaring a historiographic allegiance.[67] The engravings of the coins, set off from the rest of the book, and Lastanosa's humble claims about his inability to make sense of their inscriptions, attest to the influence of the modest or naked style. At the same time, Lastanosa cannot resist the grand gesture of the sublime. The descriptions show his deep concern with the propriety of the story the coins will one day tell. There are certainly other factors at work here; Lastanosa's interest in self promotion and in adding luster to the fame of his collection of antiquities, for example. But disinterested history is the exception in every age.

Lastanosa is not alone in attempting to reconcile conflicting historiographic demands within a single work. In *La Fenix Troyana*, for example, Vicente Mares bridges the nationalist requirements of history imposed by the crown and the desire to represent the nobility of his village, Chelva, by dividing his book into sections corresponding to national and local history.[68] Chapters with titles such as "De la poblacion de España" exist alongside others that attempt to prove that the Garden of Eden was in fact located in Chelva.[69] This kind of strategy was commonplace. What is particularly helpful about Lastanosa's book, however, is the way in which it sifts apart and then juxtaposes the conflicting tendencies in Baroque historiography into, on the one hand, a graphic, naked series of engravings, and on the other, their corresponding texts (a word that in its very etymology—from *texere*, "to weave"—suggests clothing). While the *Museo* cannot be numbered among the great works of Baroque history, few other books so clearly grapple with the broad range of representational issues facing historians during the seventeenth century. In the final analysis, Lastanosa's work is itself illuminating precisely because it fulfils the promise made in the epigraph: to cast light on nudity.[70]

Notes

1. Gaius Plinius Secundus, *C. Plinii Secundi Naturalis historia* (Hildesheim and New York, Georg Olms, 1992), *Praefatio* 15.
2. José Vicente del Olmo (1680), *Relacion historica del auto general de Fe, que se celebro en Madrid Este Año de 1680...*, ([Madrid], Por Roque Rico de Miranda, 1680), "Aprobación."
3. Juan Francisco Andrés de Uztarroz, *Vida de San Orencio, Obispo de Aux...* (Zaragoza, por Pedro Lanaja y Lamarca, 1648). In the dedicatory "A los ilustrissimos Señores...," Uztarroz explains that a historian's hard work and good judgement "caminos ignorados descubren, empresas arduas desprecian," and "embaraços dificultosos vencen."
4. Vincencio Juan de Lastanosa, *Museo de las medallas desconocidas españolas...*, Huesca, Por Iuan Nogues, 1645.
5. Lastanosa, *Museo*, "A los Lectores."
6. Lastanosa, *Museo*, 53.
7. Lastanosa, *Museo*, "Prefacion."
8. Jerónimo de Huerta, *Historia natural*, (Madrid, Por Luis Sánchez, 1624), 3. A comparison of Lastanosa's epigraph and Huerta's translation indicates that Lastanosa at least consulted Huerta's well-known work. Huerta's translation reads: "Cosa es dificil dar novedad a las cosas viejas, autoridad a las nuevas, lustre a las desusadas, luz a las escuras, gracia a las desgraciadas, y fe a las dudosas; pero mas dificil es dar su naturaleza a todas las cosas, y dar a su naturaleza todas las

propiedades secretas. Y assi quando esto no se pueda alcançar, o hazer; solo aver querido, es obra grandemente magnifica y bella."

9. The debate over whether Clio should be clothed or naked is still lively during the nineteenth century. See Stephen Bann, *The Clothing of Clio: A study of the representation of history in nineteenth-century Britain and France*, (Cambridge and New York, Cambridge University Press, 1984).

10. For a general introduction to Cabrera's historiographic ideas, see Harry Sieber (1999), "Teoría y práctica del discurso historiográfico: *Felipe II, Rey de España* (historia escrita por Luis Cabrera de Córdoba)," *Edad de Oro* 18 (1999): 207-14. On San José, see Gonzalo Fontana Elboj, "El *Genio de la Historia* de Fray Jerónimo de San José en el marco de la tratadística histórica del Humanismo," *Alazet* 14 (2002): 139-56; and María Teresa Cacho Palomar, "Una poética para una escuela: el *Genio de la Historia*," in: José Enrique Gil Laplana, ed., *La cultura del Barroco. Los jardines: arquitectura, simbolismo y literatura (Actas del I y II Congreso en torno a Lastanosa)*, (Huesca, IEA, 2000), 15-28. Whereas Fontana Elboj focuses on the debt San José owes to classical treatises, Cacho Palomar locates the *Genio de la historia* within wider considerations of textual representation carried out in Aragón during the seventeenth century.

11. John Slater, "History as an Ekphrastic Genre in Early Modern Spain" *MLN*, 122.2 (2007): 216-32.

12. Luis Cabrera de Córdoba, *De Historia, para entenderla y escrivirla*, (Madrid, Por Luis Sánchez, 1611), 5. According to the *Catálogo*, Lastanosa owned this and many of the books discussed in this article.

13. In his "Aprobación" to Fabro Bremundan's *Viage del Rey Nuestro Señor D. Carlos II al Reyno de Aragon* (1680), Manuel de Guerra y Rivera complains that among historians, some "escriven pagados: otros quejosos: otros apasionados. Pocos indiferentes, y todos hombres. O pobres verdades, que vivis entre tales cortinas!" Images that portrayed the truth as obscured by curtains or clothing were frequent during the period.

14. Gonzalo de Céspedes y Meneses, *Historia apologetica en los sucessos del Reyno de Aragon y su Ciudad de Çaragoça, años de 91 y 92...*, (Zaragoza, Por Iuan de Lanaja y Quartanet, 1622), 2, emphasis added.

15. Francisco Pinel y Monroy, *Retrato del buen vassallo copiado de la vida, y hechos de D. Andres de Cabrera, Primero Marques de Moya...*, (Madrid, En la Imprenta Imperial, por Ioseph Fernandez de Buendia, 1677), "Dedicatoria."

16. On the "metatexto historiográfico," see Walter D Mignolo, "El Metatexto Historiográfico y la Historiografía Indiana," *MLN* 96 (1981): 358-402.

17. Antonio de Solis y Rivadeneyra, *Historia de la conquista de Mexico...*, (Barcelona, en la imprenta de Joseph Llopis..., 1691), "A Los que Leyeren."

18. Olmo, "Aprobación."

19. Juan Cortés Osorio, *Constancia de la Fee, y Aliento de la nobleza española...*, (Madrid, En la Imprenta de Antonio Roman, 1684), 366.

20. Cacho Palomar, "Una poética para una escuela", 25, studies matter and form in San José's *Genio de la historia.*

21. Jerónimo de San José, *Genio de la historia*, (Zaragoza, En la Imprenta de Diego Dormer, 1651), "Advertencia del autor, al que publicare este tratado." On

the importance of the absence of marginal notes in the *Genio de la historia*, see Aurora Egido, *Las caras de la prudencia y Baltasar Gracián*, (Madrid, Castalia, 2000), 180.

22. Catherine of Sienna, *Ramillete de epistolas y oraciones celestiales para fecundar todo genero de Espiritus...*, (Barcelona, A costa de Juan Cassañes y Iayme Surià, 1698), "Parecer."

23. Alfonso Sánchez, *De rebus Hispaniae Anacephalaeosis libri septem...*, (Compluti, Typis Antonii Duplastre, 1634), 1-2 This book was also owned by Lastanosa.

24. Francisco Fabro Bremundan, *Floro Historico de la guerra sagrada contra turcos. Segunda parte...*, (Madrid, En la Imprenta de Antonio Roman, 1686). On the relationship of Fabro Bremundan and Lastanosa, see Carlos Garcés Manau, "Vincencio Juan de Lastanosa: una biografía," in *Vincencio Juan de Lastanosa (1607-1681): La pasión de saber*, (Huesca, IEA, 2007), 36; and Ricardo del Arco Garay, *La erudición aragonesa en el siglo XVII en torno a Lastanosa*, (Madrid, Cuerpo facultativo de archiveros, bibliotecarios y arqueólogos, 1934), 142.

25. Cristóbal Suárez de Figueroa, *Historia y anal relacion de las cosas que hizieron los Padres de la Compañia de Jesus...*, (Madrid, En la Imprenta Real, 1614), "Al lector." The *Historia y anal relacion* presents a particularly interesting case, because it is a translation of Guerreiro's *Relacam annal* (1609). Cabrera praises his friend Suárez de Figueroa not for the information the book contains—that is Guerreiro's work—but for the style in which this information is preserved.

26. Cabrera de Córdoba, *De Historia*, 61.

27. It is probable Cabrera's disagreements with Bartolomé Leonardo de Argensola, who was ordered by Philip III to censor the second part of Cabrera's biography of Philip II, prevents its publication. Argensola was highly regarded by Lastanosa and his friends and the dispute with Cabrera may have affected how the *De historia* was read in Aragón. On Cabrera and Argensola, see: Santiago Montero Díaz, "Estudio Preliminar," in: Luis Cabrera de Córdoba, *De historia, para entenderla y escribirla*, (Madrid, Instituto de Estudios Políticos, 1948), xxxix. The tensions between Aragonese and Castilian historians—including Cabrera, among the Castilians and Céspedes y Meneses, among the Aragonese—are studied by Richard L. Kagan, "Clio and the Crown: Writing History in Habsburg Spain," in: Richard L. Kagan and Geoffrey Parker, eds., *Spain, Europe and the Atlantic World*, (Cambridge and New York, Cambridge University Press, 1995), 73-99.

28. Juan Briz Martínez, *Historia de la fundacion, y antiguedades de San Iuan de la Peña...*, (Zaragoza, Por Iuan de Lanaja y Quartanet, 1620), "Prólogo al lector," emphasis added. Lastanosa, *Museo*, 57-8, cites this work on pp. 57-8. Del Arco, *La erudición aragonesa*, 142, reprints an example of the correspondence between Briz and Lastanosa that attests to their friendship.

29. On the friendship of Torre y Sevil and Lastanosa, see Manuel Alvar, *Edición y estudio del* Entretenimiento de las musas *de Don Francisco de la Torre y Sevil*, (Valencia, Universitat de València, 1987), especially 7-8.

30. Francisco de la Torre y Sevil [Sebil], *Luzes de la Aurora, dias del sol, en fiestas de la que es sol de los dias, y aurora de las luzes, Maria Santissima...*, (Valencia, por Geronimo Vilagrasa, 1665), "Prólogo."

31. On San José's distaste for the sublime style see Egido, *Las caras*, 157-89.
32. Cacho Palomar, "Una poética para una escuela", 24-26, points out that San José and Uztarroz are both opposed to extremes, whether of brevity or sublimity.
33. Juan Francisco Andrés de Uztarroz, *Progressos de la historia en el Reyno de Aragon, y elogios de Geronimo Zurita, su primer coronista...*, (Zaragoza, Por los Herederos de Diego Dormer, 1680), 226.
34. San José, *Genio de la historia*, 147.
35. Cabrera de Córdoba, *De Historia*, 13.
36. Cabrera de Córdoba, *De Historia*, 12, 15.
37. Cabrera de Córdoba, *De Historia*, 67.
38. José Martínez Millán and Carlos Javier de Morales, "Introducción," in: Luis Cabrera de Córdoba, *Historia de Felipe II, Rey de España*, 4 vols, (Salamanca, Junta de Castilla y León, 1998), xvii.
39. Antonio González de Rosende, *Vida i virtudes del Ill[ustrissimo] i Exc[elentissimo] Señor Don Iuan de Palafox i Mendoza...*, (Madrid, por Iulian de Parede, 1666), "Censura."
40. Baltasar Gracián, *Obras completas*, (Madrid, Espasa Calpe, 2001), "Censura", 1014-15. Uztarroz even finds it necessary to defend the length of his own *censura*, stating, "El cariño que tengo a estos escritores que se censuran en este escrito me ha dado ocasión de dilatarme más de lo que permite la brevedad de una aprobación."
41. More than out of fondness for Cabrera, it seems probable that Uztarroz's interest in the expansive style owes more to his admiration for Jerónimo Zurita, who was claimed by many Aragonese historians as an intellectual forbear. Uztarroz defends Zurita's "estilo copioso" in *Progressos de la historia*, dedicating two chapters to the subject: in the sixteenth chapter of the second book, "Escrive Geronimo Zurita sus anales en el estilo copioso, conforme lo que enseña Plinio Cecilio Segundo"; and the third chapter of the third book, "Inclinacion de Geronimo Zurita a los escritos de Cornelio Tacito ... aunque no le imita en lo breve del estilo, le sigue en la prudencia y claridad." On the possibility that that this defense of Zurita's expansive style formed part of a longer treatise on style, now lost, entitled "Estilo copioso defendido" see Ricardo del Arco y Garay, *La erudición española en el siglo XVII*, 2 vols, (Madrid, Instituto Jerónimo Zurita (CSIC), 1950), 869.
42. Tomás Tamayo de Vargas, *Restauracion de la ciudad del Salvador i Baia de todos sanctos en la provincia del Brasil...*, (Madrid, por la Viuda de Alonso Martin, 1628), "Dedicatoria."
43. Like propriety, rhetorical decorum is a means of relating the status of the historical subject to the nature of its representation in history. See Anthony Grafton, *What Was History? The Art of History in Early Modern Europe*, (Cambridge and New York, Cambridge University Press, 2007), 34-38.
44. San José, *Genio de la historia*, 71.
45. San José, *Genio de la historia*, 150. Egido, *Las caras*, 180 briefly discusses modesty and the "verdad desnuda" in San José.
46. San José, *Genio de la historia*, 162.
47. San José, *Genio de la historia*, 147-48.

48. San José, *Genio de la historia*, 147.
49. Egido, *Las caras*, 178-79 and 187. Egido hints at the fact that it is not always clear how verisimilitude—the historian's proper handling of narrative—is to be distinguished from the historian's claim to authority in the *Genio de la historia.*
50. Scholars have tended to pay more attention to the systems for establishing historical authority in the *crónicas* of the Spanish colonies. It is important to note that the tenor and substance of the debates surrounding historical authority in the colonies and on the peninsula were very different. For example, there is little focus on first-hand witness among peninsular historians who instead spend much more time defending philological and archival practices.
51. Gianna, Pomata and Nancy G. Siraisi, *Historia: Empiricism and Erudition in Early Modern Europe*, (Cambridge, Mass., MIT Press, 1996), 1-38. Pomata and Siraisi explain that early modern "Historia"—a term they use in English to describe a wide variety of documentary activities—is an epistemic, as well as literary, genre.
52. On Lastanosa's goals for the *Museo*, see Aurora Egido, *La rosa del silencio: Estudios sobre Gracián*, (Madrid, Alianza, 1996), 190-96.
53. Pinel y Monroy, *Retrato*, "Dedicatoria."
54. Lastanosa, *Museo*, 54.
55. In the "ilustraciones" that follow, Rajas' arguments—which "si no convencen, hazen la resolucion muy verisimil"—foreground the plausibility of the narration over the authority of the historian. See Lastanosa, *Museo*, 118, 121.
56. Lastanosa, *Museo*, 54, 56.
57. Lastanosa, *Museo*, 53-116
58. Lastanosa, *Museo*, 54.
59. Lastanosa, *Museo*, 53.
60. The "Advertencias" sometimes fail to correspond to the coins altogether. See Almudena Domínguez Arranz, "Monedas, medallas y piedras preciosas en el 'museo discreto' de Vincencio Juan de Lastanosa," in: *Vincencio Juan de Lastanosa (1607-1681): La pasión de saber*, (Huesca, IEA, 2007), 126.
61. Karl Ludwig Selig, "Góngora and Numismatics" *MLN*, 47-50 (1952): 48.
62. Lastanosa, *Museo*, 59-61, 103.
63. Egido, *La rosa del silencio*, 193. Especially helpful for understanding the relationship between debates about poetry and those related to historiography is Cacho Palomar, "Una poética para una escuela", and Aurora Egido, *La poesia aragonesa del siglo XVII (Raíces Culteranas)*, (Zaragoza, CSIC, 1979), 24-36. Too often, historiographic debates have been understood to be derivative of their poetic counterparts. This is an unfortunate and limiting error. Cacho Palomar and Egido show that in Aragón during the seventeenth century we find a broad interest in representation not limited to or privileging any particular genre, rather than historiographic concerns being subordinated to poetic.
64. Lastanosa, *Museo*, "A los lectores." As Kagan, "Clio and the Crown", 83 points out, the very year that the *Museo* is published, 1645, Philip IV institutes new laws to see that even local histories serve royal ends. Lastanosa's work neatly straddles this requirement, claiming to honor both the author's home town of Huesca and the nation.

65. Lastanosa's desire to recuperate the lost history both of his native Aragón and of Spain can be seen as part a collective project undertaken by many in his circle. On Uztarroz's interest to "restituir a Aragón su verdadera historia" see Egido, *Las caras de la prudencia*, 159-60.

66. Egido, *La rosa del silencio*, 191 points out that Lastanosa, unlike his friend Gracián, believed himself to be living "en un siglo feliz."

67. Gracián, *Obras completas*, 1094.

68. On the nationalist demands placed upon local histories see Kagan, "Clio and the Crown", 84-99.

69. Vicente Mares, *La Fenix Troyana. Epitome de varias, y selectas historias, assi Divinas, como Humanas: Breve resumen de la poblacion del universo...*, (Valencia, Por Mateo Penen, 1681).

70. I would like to thank James Mandrell, Andrés Prieto, Aurora Egido, Jorge Cañizares Esguerra and María Luz López Terrada for their helpful suggestions at various stages of writing.

CHAPTER FOUR

TYPOLOGICAL READINGS OF NATURE: THE BOOK OF NATURE IN LASTANOSA'S AGE

JORGE CAÑIZARES-ESGUERRA

The image of a Franciscan mission in the New World in Diego Valadés' *Rhetorica christiana* (Perugia, 1579) typifies early-modern Spanish clerical ideas on conversion.In addition to using images (*creation mundi*), rebus writing (*discunt omnia*), or simply books (*discunt doctrinã; discunt penitentiã; discunt confiteri*), the Franciscans use nature to introduce the natives to the Bible or the sacraments of the church. A tree is displayed like an illuminated manuscript to teach the natives a lesson about the sacrament of matrimony (*examen matrimonii*) (Fig. 1).

There is nothing surprising about the Franciscan use of a tree to teach the natives a religious lesson. From its very beginnings Christianity assumed that God spoke to humankind through two different books: the Bible and Nature. The study of the alphabet and grammar of the Book of Nature revealed, in the words of Luis de Granada, a Dominican friar contemporary of Valades and Lastanosa, the "generosity, wisdom, and omnipotence" of the Lord.[1] According to Granada the fixed stars and the planets; the spheres of fire, air, water, and earth of the sublunary world; the creatures of the earth (fish, quadrupeds, birds, and insects); and the structure of the human body were all texts that spoke of God's power, lawfulness, and generosity. Every structure in nature had been created to serve and teach humanity a lesson. The flora and fauna were there to feed, heal, and shelter humankind. Bees not only provided humans with honey and wax but taught lessons about the appropriate structure of the human polity.[2] Flies, fleas, ants, and spiders, seemingly on earth with no purpose other than to annoy humans, had in fact been created to testify to the awesome microscopic designing power of the Lord.[3] Even snakes, dragons, tarantulas, and other poisonous creatures had a purpose: to punish

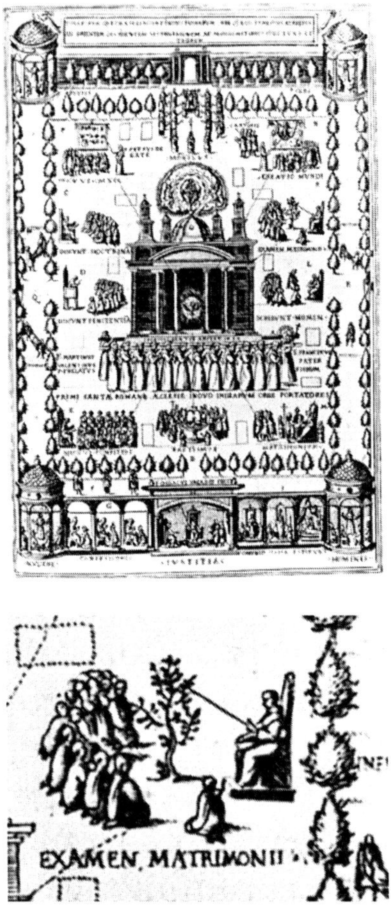

Fig. 1. Diego Valadés' *Rhetorica christiana* (Perugia, 1579). Author's collection.

and make miserable the lives of the descendants of Adam and Eve after the Fall.[5]

But Valades' image reveals other aspects of Franciscan culture that historians of science have overlooked when studying medieval and early modern European interpretations of nature. In addition to describing how a typical Mexican mission worked, Valades presents the Franciscan pastoral expansion into the New World as being foreordained in the Bible. Led by Martin Valencia and St. Francis himself, twelve Franciscans carry the Holy Roman Church into the New World for the first time ("*primi sanctae romane aeclesie [sic] in novo indiarum orbe portatores*"). In Genesis 28:14-15 the Lord has already prefigured their task: "And thy seed shall be as the dust of the earth, and thou shalt spread abroad to the west, and to the east, and to the north, and to the south... And, behold, I am with thee, and will keep thee in all places whither thou goest" (*dilatarebis ad orientem, occidentem, septentrionem, ac meridiem et ero custos tuus et tuorum*). The Church the twelve Franciscans carry is inhabited by the Holy Spirit (*spiritus santus abitat in ea*), corresponding to the millennial, third spiritual age of Joachim de Fiore. The Church in the Novo Indiarum Orbe is a walled garden, a New Jerusalem: its walls rest on twelve stone foundations bearing the names of the twelve apostles of the Lamb (Rev. 21:14) (the twelve small rectangles scattered on the ground seem to be both the tombs of the original Franciscan apostles and the stone foundations of the new millennial Church). The Holy Spirit that descends from the throne of God fertilizes a path of trees where the infirm (*infirmi*) are taken to be healed. The trees in the Franciscan compound thus seem to correspond to the Trees of Life of the New Jerusalem whose leaves are "used as medicine to heal the nations" (Rev. 22:2). In this New Jerusalem, the friars mete out justice, bury the dead, perform the sacraments (confession, communion, last rights, baptism, matrimony, penance, and the Eucharist), and teach the Indians music and to read and write.[6]

Typology is an ancient reading technique that presents newer events as fulfillment of older ones. The New Testament itself originated as a typological text, as the Gospel writers strove to cast the life and deeds of Mary, the Apostles, and Christ as the fulfillment of Old Testament types. In the hands of, say. Luke and Mark, Christ became the fulfillment of Moses and David. In patristic times, typology became a reading technique to interpret the history of the church itself as the fulfillment of both Old and New Testament texts. Augustine, for example, interpreted the book of Revelation as a prefiguration of the history of the church, leading to the final triumph of the city of God. For Augustine the biblical account of creation in seven days prefigured the history of humanity in seven ages. In the Middle Ages, the Calabrian abbot Joachim di Fiore transformed typology into an art and divided human history into three stages, each a prefiguration of a person of the Trinity: the father (Old Testament), the son (New Testament) and the Holy Spirit (new millenarian dispensation). In the Middle Ages, typology led to the development of the Marian litanies, a long list of Old Testament figures prefiguring the virtues and piety of the Virgin: Ark of God, Pillar of Fire, Burning Bush, Rod of Jesse, Tower of David, Rose of Sharon, Hortus Conclusus.[7] (Fig. 2)

Fig. 2 Attributed to Jerónimo del Río Dieste (1730-1732).. Marian Symbol (Tower of David) in Church of St. Dominic. Huesca. From Fernando Alviria Lizano and Maria Celia Fontana Calvo, *La Iglesia Oscence de Santo Domingo* (Huesca: Instituto de Estudios Altoaragoneses, 2006)

But typology as a reading technique was not restricted to the Bible. It was also useful for interpreting classical sources. Under typological scrutiny, Greek and Roman mythology became a prefiguration of Old and New Testament narratives. In the Middle Ages, Theodolphus's *Ecloga* (c.800), Bernard of Utrecht's *Kommentar* (late 13th c) and Petrus Berchorius' *Ovidius moralizatus* (c 1350), for example, transformed every ancient myth into a biblical type. When Juan Vicencio de Lastanosa (1607-1684) was busily assembling his library and cabinet of curiosities, Filippo Picinelli wrote a biblical commentary based on Greek and Roman myths.[8] Most likely a fixture of Lastanosa's library, Picinelli's *Mundus Symbolicus* (1653) was one of the most important encyclopedias of emblems in the Catholic world in the seventeenth century.[9] To be sure, Picinelli drew on typological readings of both Old Testament figures and Greek and Roman mythology to create his emblems on Catholic virtues. But more important, Picinelli also drew on typology to interpret the Book of Nature. Thus a quick glance at this encyclopedia shows that every element in the cosmos and every creature on earth had been designed by the Creator as a prefiguration of the narrative of Christian salvation. This is the world that Valades and Lastanosa inhabited, one held together by typological readings of the Bible, classical mythology, and the Book of Nature.

Like Picinelli, Lastanosa associated Christian virtues to Old Testament types and Classical myths, as descriptions of the paintings in his palace and gardens seem to attest. But like Picinelli, Lastanosa also read nature typologically. Lastanosa's emblem attached to his *Museo de las monedas desconocidas españolas* (Huesca, 1645) shows a Phoenix under the motto "Vetustate fulget" (shines in old age) (Fig. 3). We do not know why Lastanosa represented himself as a Phoenix in old age. Yet it is clear that in the early-modern Catholic circles of Lastanosa and Picinelli, the phoenix was no mere metaphor for bodily resurrection: It was a real bird somewhere in Arabia that reportedly buried itself in scented piles of frankincense and myrrh as it grew old, only to be born again out of a winged worm in the original decomposing corpse. If one believed that the Book of Nature introduced humans to the omnipotence, wisdom, and generosity of the Creator, what lessons would one draw from the life cycle of the phoenix? According to Luis de Granada, the phoenix had been created by God to prefigure the resurrection of the body of the saints.[10] Like the Bible and classical sources, Nature was a text that could be read typologically. Granada's *Introducción del Símbolo de la Fe* [1583] offers countless examples of such readings.

Take for example Granada's interpretation of the sun. After a detailed description of planetary motions and their seemingly changeless regularities, Granada studies the sun: its brilliant glow is such that at night all stars, including the moon, reflect its light; its penetrating virtue reaches deep inside the earth's bowels, causing gold and precious stones to grow; its obvious visibility contrasts with the difficulty of seeing it directly, for it burns the eye; its heat penetrates through doors and walls even if one seeks to take shelter; it sets off all motion and change of elements in the sublunary sphere; finally, it is the source of both light and darkness. According to Granada, God designed the sun as a prefiguration of His own virtues and powers: God's glow is everywhere and it reflects off the deeds of the saints; He causes spiritual riches to grow deep inside fallen human nature; He is obviously visible in creation yet remains unintelligible; He is the First Cause of everything in creation; finally His brilliant presence leads to illumination and his absence to darkness and evil. Using the same terms used in patristic theology to refer to the Old Testament as a "shadow" (anticipation) of the New Testament, Granada concluded that all creatures in Nature were also "little shadows or marks of the Lord" (*una pequeña sombra o huella del Criador*).[11]

Fig. 3. Lastanosa's Emblem. In Vicencio Juan de Lastanosa, *Museo de las monedas desconocidas españolas* (Huesca, 1645) Instituto de Estudios Altoaragoneses.

Granada offered many other examples of nature as "little shadows" of God. Why would the sea be pleasantly calm at times and furiously agitated at others? It is because the Lord prefigured in the changes of the ocean both His tenderness for the saintly and His wrathfulness for the sinful (219-220). The Lord designed the grapevine in a trellis as a prefiguration of the way humans achieve salvation by crawling their way up the cedar tree that was Christ (249); the castrated rooster (capon) grows fat because it anticipates the spiritual plumpness that awaits the celibate (276); the hearty appetite of lambs getting ready for winter prefigures the behavior of the saintly who binge in good works not to be deprived of merit in the afterlife (275); the burns that develop on the wings of the stork when she shelters her chicks from the scorching rays of the sun and the self-inflicting chest wounds of the pelican to feed her young with her blood anticipate the Passion of Christ to save His offspring (289, 333); the instinct of quails to follow their mothers after hatching, even after the eggs had been stolen by rival females, anticipates the conversion of the pagans to Christianity, for the latter were also capable of identifying their true spiritual parents (332). The list continues endlessly. Granada even describes the structure, routine, hierarchies, and patterns of discipline of the beehive as a prefiguration of an idealized monastery (355-66).

The typological, prefigurative reading of the Book of Nature as the case of Granada suggests had deep roots in the bestiaries of the Middle Ages which found the beaver's self-castration to evade the hunter to be a prefiguration of celibacy to avoid sin, the female elephant's seduction of a reluctant male elephant (using the fruit of the mandrake) to be a figure of Adam and Eve, and the unicorn's surrender to a maiden to represent the Passion of Christ.[12] In the emblematic literature of the early modern period these typological readings became political advice. Granada's reading of the pomegranate, again, is typical of the genre. The pomegranate, a fruit made of many countless small red grains, had long stood for peaceful cohabitation: In Cesare Ripa's Iconologia (1593) it represented concord (emblem 56). In the eighteenth century edition of Ripa, however, the pomegranate now took the meaning of "democracy" as all the seeds were thought to be similar and equal (fig. 4).

But for Granada, in addition to its many seeds, the pomegranate had a crown-like structure on top. According to Granada, the pomegranate not only was "the queen of all fruits" but also the original inspiration for the crown of kings (248). In Granada and Lastanosa's age the pomegranate was thought to prefigure the ideal polity, a harmonious unity of multiple kingdoms: the composite monarchy. In 1519, Albrecht Durer chose to portray the Holy Roman Emperor Maximilian I, the father of Charles V, holding a pomegranate..

Fig. 4. Pomegranate as prefiguration of Democracy. In Cesare Ripa *Iconologia* (ed. C. Orlandi and R. Sangro, Perugia, 1764-67). Author's collection.

The typological reading of Nature to draw political lessons reached its apogee in the early modern literature on emblems, a genre that we know Lastanosa fondly cultivated. Again Picinelli's *Mundus Symbolicus* is typical of this literature, for Picinelli goes over the forms and virtues of stars, planets, elements, birds, quadrupeds, fish, poisonous animals, insects, trees, fruits, herbs, flowers, rocks, stones, and metals (in addition, of course, to the Old and New Testaments and Greek and Roman mythology), and in so doing brings forth either narratives of Christian salvation or advise for the prince and the virtuous to follow. In Spain the literature on emblems was no different. Take for example the case of Diego Saavedra Fajardo's *Idea de un principe politico christiano representada en cien empressas* (1642).

Saavedra's emblems draw repeatedly upon nature. Unlike Granada who took the sun to be a type of God, Saavedra assumes it to be a prefiguration of kingly authority. The reflecting glow of the stars and the moon, for example, becomes a prefiguration of the appropriate subordination and quality of *validos*, for the authority and prestige of courtiers is a reflection of the glow of the sun, the king (empresa 49) (Fig. 5)

Fig. 5. Courtiers as reflections of the Sun Monarch (emblem 49). In Diego Saavedra Fajardo's *Idea de un principe politico christiano representada en cien empressas* (1642). Author's collection.

The nature of solar eclipses teaches kings not to communicate with other princes directly but through ministers, for the greater the proximity of an object to the sun the greater the shadows: rivalries and misunderstandings (empresa 77). Lunar eclipses, on the other hand, teach the prince that his own authority is a reflection of that of God, the sun (empresa 13). Night is the shadow cast upon the the earth by the absence of the sun, and reminds the prince to shine forthrightly so as to expel darkness and malice from the court (empresa 12). The annual transit of the sun through the ecliptic (zodiac) prefigures the rounds of the king through his kingdoms (empresa 86). The illuminating power of the sun prefigures the sovereignty of the Pope over the earth as a whole (empresa 94). The magnetic north and the polar star show the need for the king to pilot the ship of state unwaveringly, always returning to the same port of call: God (empresa 24; see also empresa 36). Saavedra draws on the behavior of rivers (empresa 90), water (empresa 65), mountains (empresas 40 and 50), volcanoes (empresa 93), magnets (empresa 88), trees (empresas 54, 67, and 70), vines (empresas 66 and 74), palms (empresa 96), flowers (empresas 6 and 87), wheat (empresa 6), roses (empresa 34), corals (empresa 3), oysters (empresa 32), bees (empresas 42, 62, 73, 85, and 99), storks (empresa 25), parrots (empresa 79), ravens (*corneja*) (empresa 47), ostriches (empresa 22), snakes (empresas 28, 44, and 79), salamanders (empresa 48), scorpions (empresa 52), eagles (empresa 22 and 92), falcons (empresa 10), lions (empresas 33, 45, 97, and 99), horses (empresa 38), dogs (empresa 9), porcupines (empresas 59 and 82), bulls (empresa 80), bears (empresa

85), and unicorns (empresa 8) to draw the appropriate advice for the prince. Beyond the huge literature on emblems, the study of animals and plants to draw religious and political insight was a constant in seventeenth century Spain.[13]

What tree the Franciscans might have used to teach the natives of Mexico lessons on the sacrament of matrimony is difficult to tell. Was it perhaps a myrtle (a symbol of fertility and fidelity in classical sources and of prophetic hope and Marian purity in the Old and New Testaments) or a poplar (a tree with funerary connotation but also of marriage given the attraction of the grapevine for it)?[14] The use of flowers and trees as types of Christian narratives was well established in ancient and medieval Europe. In the Middle Ages, botany studied the medical properties of plants but also the connections of flowers to the life and virtues of the Virgin Mary. What possible other function could the extraordinary array of flowers have for humans besides being a source of aesthetic pleasure, scents and oils?[15] The answer usually came down to their prefigurative ability to speak of the life of Christ and the Virgin. Since its corolla faces down and it appears along with the rose of Sharon as one of the flowers that describe the female lover in the Song of Songs (2:1), the lily of the valley came to represent both the humility and purity of Mary. The columbine with its dove-like petals and six flowers per stem prefigured the Holy Spirit and His six gifts: wisdom, intelligence, counsel, fortitude, knowledge, and fear of the Lord (Isaiah 11:2). The intensely red poppy, which grows amid fields of wheat and whose scent induces drowsiness, was a type of the Eucharist and of the Passion of Christ.[16]

The list is endless. Suffice it to say that entire "gardens of Mary" surfaced in monasteries and nunneries to grow flowers that prefigured the Annunciation (Mary lily or *Lilium candidum*; Our Lady's Modesty or *Viola odorata*), the Visitation (Columbine or *Aquilegia vulgaris*), the Nativity (Our Lady's Bedstraw or *Galium verum*; Mary's Love of God or *Dianthus caryphyllus*; Christmas Rose or *Helleborus niger*; Mary's Star or *Chrysanthemum leucanthemum*; Star of Bethlehem or *Ornithogalum umbellatum*; Our Lady's Thistle or *Carduus benedictus*), the Presentation (Flower of the Purification or *Galanthus nivalis*; Mary's Tears or *Pulmonaria officinalis*), the Flight into Egypt (Virgin's Bower or *Clematis vitalba*; Saint Mary's Tree or *Rosemarinus officinalis*; Juniper or *Juniperis*; Our Lady's Resting Place or *Veronica chamaedrys*; Rose of Jericho or *Anastatica hierochuntica*; Our Lady's Cushion or *Armeria maritime*), the Maternity of Mary (Eyes of Mary or *Myosotis scorpioides*; Our Lady's Slipper or *Cypripedium calceolus*; Our Lady's Eardrop or *Fuchsia magellanica*; Fruitfulness of Mary or *Fragaria vesca*), Mary the

Homemaker (Our Lady's Smock or *Cardamine pratensis*; Our Lady's Thimble or *Campanula rotundifolia*; Mary's Drying Plant or *Lavandula officinalis*; Mary's God or *Calendula officinalis*), Mary at the Cross (Mary's Tears or *Convallaria majalis*), and Mary's Tomb (Red Rose; Mary's Lily or *Lilium*).[17] (fig. 6).

In Lastanosa's age, Calderon de la Barca was perhaps the most accomplished typological interpreter of nature. Calderon left behind numerous *autos sacramentales* (plays for Corpus Christi) based on figural readings of the Old Testament (*El cordero de Isaías*; *El primero y segundo Isaac*; *La piel de Gedeón; El arca de Dios cautiva; Las espigas de Ruth; El nuevo palacio del Retiro* [on the Garden of Eden and Noah's Ark as prefigurations of the new palace of El Retiro]) and of Greek and Roman myths (*El divino Jasón*; *Andrómeda y Perseo*; *El divino Orfeo; El verdadero Dios Pan; El pastor Fido; El sacro Parnaso; Los encantos de la culpa* [on Circe and Ulysses]). To be sure, he also wrote *autos* drawing on the Book of Nature. In *El lirio y la azucena* (1660), Calderon deploys the natural, prophetic, and historical associations of the iris and lily to celebrate the Treaty of the Pyrenees of 1660 between France (the iris of Clovis, the legendary first Frankish king) and Spain (the lily of Rudolph I, the founder of the Hapsburg dynasty). He offers a typological interpretation of the wedding of the Spanish infanta Maria Teresa and the French king Louis XIV: the union of the bride of the Song of Songs, the lily, to Christ (Louis XIV, the iris). But it is in *La humildad coronada de las plantas* [The crowning of the most humble of plants] (1644) where Calderon offers his most complex and lengthy typological reading of trees and plants.

The play has characters such as the Espino (thistle), Cedro, Encina (holm oak), Laurel, Moral (mulberry tree), Olivo, Almendro, Espiga (wheat), and Vid (grapevine) competing with one another over who should be crowned as the most valuable of plants. The cedar stands for the Cross and it is actually made of three different types of wood (cypress, palm tree, and cedar proper), a reference to the Trinitarian nature of Christ. Since the cedar is so obviously the winner, it does not participate; it judges the competition. The thistle, which has beautiful flowers but also thorns, stands for Judaism and its punitive law. The holm oak, producer of acorns (*bellotas*), appropriate for feeding pigs, stands for heresy. For its rush to blossom, the almond prefigures carelessness and madness (locura). The mulberry tree, on the other hand, is the type of prudence for its fruit mature

Fig. 6. Jan van Kessel. Wreath for the Virgin and Baby Jesus (ca. 1650). Felix Palacios Remondo Collection. Zaragoza.

slowly, feeding the silkworm. Olive is the type of the primitive Golden Age, a time of peace when Natural Law ruled. Laurel prefigures paganism, Roman and Greek antiquity, an age of honor and polytheism. The winners of the competition turn out to be the grapevine and the wheat, the types of the blood and body of Christ, a prefiguration of the Eucharist. Calderon uses the properties of trees to offer a complex interpretation of Judaism (Mosaic law), paganism, apostasy, natural law, transubstantiation, the Trinity, and political prudence.

By the time Lastanosa was designing his gardens and assembling his cabinet of curiosities, the typological reading of nature was already incorporating the curiosities of the New World. The American "passion flower" soon became a quintessential Catholic flower (fig. 7).

According to the Spanish Jesuit Juan Eusebio Nieremberg, a contemporary of Lastanosa, the flower clearly represented the whip and the stump, Christ's red wounds and blood, the three nails of the Cross, the crown of thorns, and the lance that had penetrated Christ's torso. The plant also bore the "fruit" of resurrection, delicious with sweet nectar. Both the flower and the fruit, according to Nieremberg, were contributing to the spread of the gospel as the Pope sought to "encircle" the earth with the

Fig. 7. Juan Eusebio Nieremberg, *Historia naturae, maxime peregrinae, libri XVI* (Antwerp 1635). Author's collection.

message of the liberation brought about by the Passion and Resurrection of Christ.[20] This last point was not an exaggeration, for the Jesuits had already taken the flower and its liberating meaning into the Ottoman empire.[21]

Given Nieremberg's fascination with the occult and the allegorical in fauna and flora, it is a little surprising to find that the Jesuit did not fully explore the typological meaning of a plant that hailed from a continent long in the hands of Satan. It was for others to elaborate on this mystery. Monardes in 1574 was perhaps the first naturalist to call attention to the Christian symbolism of the passion flower.[22] Then between 1607 and 1610 several pamphlets on the meaning of the flower suddenly appeared in Italy.[23] The Jesuit Giovanni Botero (1540-1617), author of *Della ragione di stato* (1594), wrote a poem in 1607 wondering how a holy flower could have originated on a continent inhabited so distant from where the impious and inhuman Jews had committed their crime against Christ.[24]

It was the German theologian Jakob Gretser (1562-1625), however, who in 1610 first tackled the contradiction identified by Botero. After translating into the Latin Simone Parlasca's Bolognese pamphlet on the passion flower, *Ill fiore della Granadiglia* (1609), Gretser went on to offer in poetic form a theological interpretation of every detail of the flower, including the meaning of the spearlike shape of the leaves, the thorns in the petals, and why the flower seemed to have only three nails instead of the four used in the Crucifixion. (The fourth nail had been used to tack up the placard identifying Christ as King of the Jews). Since the flower itself symbolized Christ's reign in Nature, the fourth nail was superfluous. Gretser then went on to explore new territory, posing such questions as why the flower had surfaced in the Antipodes, and specifically in Peru.As heresy spread in Europe, Providence had selected America as the place for the passionflower to be discovered.[25] According to Gretser, the flower was poised to bring Peruvians back from darkness and away from their crimes and sin, making the victory of the Cross all the more impressive. Making use of a trope typical of the age, Gretser explained that the plant's lancelike leaves announced that Satan would be ousted from his kingdom by Spanish armies wielding instruments of war. The symbols of the Cross and the Passion would migrate from the New World to help put down heretical outbreaks in the Old.

With so much at stake, it is little wonder that the study of this marvelous text in Nature became a craze among learned clerics in the Atlantic Catholic world. In his 1625 description of rare plants cultivated in the botanical gardens of the Cardinal Farnese in Rome, Pietro Castelli (1570-1657) captured rather vividly the excitement that his contemporaries felt: "This plant desired by the curious on account of its great rarity; recited by the poet with divine praises; celebrated by orators with all the eloquence of the profession; investigated by the philosopher with the subtlest of reason; admiringly commended by the doctor on account of its curative power; long longed for by the sick; admired by the Theologian and revered by the pious Christian has, in short, been most wanted and sought after, and still is, by all types and nature of people on account of its mystery, power, elegant form and other marvels."[26]

Castelli was something of a theologian-philosopher himself, for in addition to recycling well-worn interpretations, this learned gardener advanced a few of his own: *inter alia*, that the white petals represented the white cloth worn by the Savior; that the hornlike tips of the white petals stood for the faith that penetrates like a sting; that the leaves of the vine with tree lobes stood for the mystery of the trinity; and that the vine and its trellis represented the need Christians had for the institutional support of the Church to ascend to heaven.

Giovanni Battista Ferrari (1584-1655), gardener for Pope Urban VIII, the same pope who sent a relic of Christ's Cross to Peru, was also taken by the flower. In his *De Florum Cultura Libri IV* (1633), the learned Ferrari, professor of rhetoric, Syriac, and Hebrew at the Jesuit college, found in the flower the lance-sponge filled with vinegar with which a Roman soldier had maliciously sought to quench Christ's thirst. Ferrari believed it very significant that the flower's fruit had the shape of the heart, "the heart [itself] of a loving God."[27] Moreover he saw as revealing of the nature of the Church that the vine's roots grew so aggressively that when confined in a jar they sent shoots through the smallest of holes down into the soil. Like Castelli, Ferrari insisted that the plant had come to Rome from the remotest corners of Mexico and Peru to inspire both poets and preachers.[28] Finally, in his *Annotationes et Additiones* to the 1651 Academy of the Lynx's edition of Francisco Hernandez natural history of Mexico, Fabio Colonna (1557-1660) maintained that the flower was ideal for keeping evil spirits away and countering sorcery.[29]

This is the world Lastanosa inhabited. His intense piety colored not only his scholarship but also the content of his collections of natural curiosities. With his brother, Juan Orencio, Lastanosa spent considerable sums of the family fortune building chapels in the cathedral and the church of St Dominic in Huesca.[30] Like good provincial intellectuals in early-modern Spain, the Lastanosas sought to provide Huesca with a glorious past in the history of Christendom. The Lastanosa avidly promoted the cults to St Orencio and St Prudencia, the alleged parents of the martyr St Laurence. Countless patriotic urban clerical intellectuals found in sixteenth and seventeenth century Spain the buried relics of apostles and proto-Christian martyrs in order to present their cities as the cradle of Christendom. It is therefore not a wild speculation to assume that Juan Vicencio Lastanosa interpreted nature typologically with the help of Biblical narratives.[31]

Notes

1. Fray Luis de Granada, *Introducción del Símbolo de la Fe* [1583], José María Balcells ed. (Madrid: Cátedra, 1989)
2. *Introducción*, 355-66.
3. *Introducción*, 335-51
5. *Introducción*, 286.
6. On the importance of typology in organizing the Franciscan missions to Mexico, see Jaime Lara, *City, Temple, Stage: Eschatological Architecture and Liturgical Theatrics in New Spain* (Notre Dame, Ind.: Notre Dame University Press, 2004).
7. On typology as an ancient, patristic, and medieval reading technique, see Leonhard Goppelt, *Typos: The Typological Interpretation of the Old Testament in the New*, trans. from the German by Donald H. Madvig (1939; repr., Grand Rapids, Mich., 1982); John J. O'Keefe and R. R. Reno, *Sanctified Vision: An Introduction to Early Christian Interpretation of the Bible* (Baltimore, Md., 2005); James

Samuel Preus, *From Shadow to Promise: Old Testament Interpretation from Augustine to the Young Luther* (Cambridge, 1969); Erich Auerbach, "Figura," in Auerbach, *Scenes from the Drama of European Literature* (New York, 1959); Friedrich Ohly, "The Cathedral as Temporal Space: On the Doumo of Siena," in Ohly, *Sensus Spiritualis: Studies in Medieval Significs and Philology of Culture*, trans. from the German by Kenneth J. Northcott (Chicago, 2005), 136–233; Richard K. Emmerson, "Figura and the Medieval Typological Imagination," in Hugh T. Keenan, ed., *Typology and English Medieval Literature* (New York, 1992).

8. *Lumi riflessi; o, Dir vogliamo concetti della Sacra Bibbia osservati ne i volumi non sacra, studii eruditi* (Milan, Stampa di F. Vigone, 1667). For similar sixteenth- and seventeenth-century studies, see Georg Witsel, *Parallela : affinia quaedam, et aliquo modo correspondentia exnostris, hoc est, Sacris & Gentilium libris* (Mainz: Franciscus Behem,1544); Johann Bompart, *Parallela sacra et profana; sive, Notae in Genesin* (Amsterdam: J. Wolters, 1689); Johannes Hoffmann and Andreas Rötel, *Deorum gentilium praecipuorum origines ex Sacra Scriptura derivatus* (Jena: Joh. Jacobi Bauhoferi, 1674); Eccardus Leichnerus, *Opuscula quae ad historiam philologiam sacram spectant* (Rotterdam: Petrum vander Slaart, 1700); Zachary Bogan, *Homerus Hebraizon: sive, Comparatio Homeri cum Scriptoribus Sacris quoad normam loquendi* (Oxford: H. Hall, 1658); James Duport, *Homeri poetarum omnium seculorum facilè principis Gnomologia, duplici parallelismo illustrata uno ex locis S. Scripturæ, quibus gnomæ Homericæ aut propè affines, aut non prorsùs ab similes* (Cambridge: Johannes Field, 1660); Jacques Hugues, *Vera historia Romana sev Origo Latii vel Italie ac Romanae vrbis e tenebris longae vetustatis in lucem producta* (Rome: Francisci Monetae,1655); Johannes Hencricus, *Ursinus, Analectorum sacrorum* (Frankfurt: Hermsdorffi, 1658-1660). On typological readings of classical myths, see Hugo Rahner, *Greek Myths and Christian Mystery*, translated from the German by Brian Battershaw (New York: Biblo, [1957] 1971); H. David Brumble, *Classical Myths and Legends in the Middle Ages and Renaissance: A Dictionary of Allegorical Meanings* (Wesport, Connecticut: Greenwood Press, 1998); and Don C. Allen, *Mysteriously Meant* (Baltimore: John Hopkins University, 1970).

9. See "Catálogo de los libros de D. Vicencio Ioan Lastanosa," Ms. U-379. Royal Library. Stocolm, published in K.L. Selig, *The Library of Vicencio Juan Lastanosa, Patron of Gracián* (Ginebra: Droz, 1960).

10. Luis de Granada, *Introduccion*, 375-76.

11. Granada, *Introducción*, 197-200.

12. *Physiologus Latinus*, 34 (elephant); 16 (unicorn)

13. See, for example: Jerónimo Cortés, *Libro y tratado de los animales terrestres y volatiles.* (Valencia 1615); Francisco Marcuello, *Primera parte de la Historia natural, y moral de las aves* (Madrid, 1617); Andrés Ferrer de Valdecebro, *Gobierno general, moral y político, hallado en las fieras y animales sylvestres* (Madrid 1658); Juan Eusebio Nieremberg, *Historia naturae, maxime peregrinae, libri XVI* (Antwerp 1635). For an importatn Portuguses text, see Antonio do Rosario, *Frutas do Brasil numa nova, e ascetica monarchia, consagrada*

a'Santissima Senhora do Rosario (Lisboa: Na Officina de Antonio Pedrozo Galram, 1702).
14. Lucia Impeluso, *Nature and its Symbols*, translated form the Italian by Stephen Sartarelli (Los Angeles: Paul Getty Museum, 2002), 55, 59.
15. Granada, *Introduccion*, 233-34.
16. Impeluso, *Nature and its Symbols*, 79, 109, 111.
17. Vicenzia Krymow, *Mary's Flowers. Gardens, Legends, and Meditations* (Cincinnati, Ohio: St Antony Messenger Press, 2002).
20. Nieremberg, *Historia Naturae*, bk 14, ch. 10, p. 300.
21. Simon Contarini to the Doge and Senate, Vigne di Pera, Oct 3, 1609 in *Calendar of State Papers and Manuscripts Relating to English Affairs Existing in the Archives and Collections of Venice and in Other Libraries of Northern Italy*, edited by Horatio F. Brown, vol XI (1607-1610), 360-61.
22. Monardes, *Primera y Segunda y Tercera Partes de la Historia Medicinal de las Cosas que se traen de nuestras Indias Occidentales (* Sevilla: Alonso Escribano, 1574), bk. 3, 109r-v.
23. Donato Rasciotti, *Copia del fiore et fruto che nasce nelle Indie Occidentali,, qual di nuovo e stato presentato all Santità di NSP Paolo V* (1609); Parlasca; Bosio.
24. Juan Botero quoted in Karel Stengel, *Hortorum, Florum, et Arborum Historia in II Tomos distributa* (Augustae Vindel: Andrea Apergeri, 1650), 370.
25. Jacob Gretser, "Descriptio Floris Indici, quem Granadillam vocant, & in eundem varia Epigramatia. Pleraque ex Italico quodam de hoc flore hypomnematio expressa," in *Hortvs S. Crvcis ... Accessit Liber de more lavandi pedes peregrinorvm et hospitvm, cvm Paralipomenis pro libris de sacris peregrinationibus & processionibus* (Ingolstad, Adami. Sartorii, 1610), quinta pars, sec. 7, 273-94.
26. Pietro Castelli, *Exactissima Descriptio Rariorum Quarundam Plantarum, Que contientur Rome in Horto Farnesiano* (Rome: Jacob Mascardi, 1625) ch. 3, p. 49.
27. Giovanni Battista Ferrari,. *Flora, overo Cultura di fiori* [1633]. Rome: Pier'Ant. Faciotti, 1638), 194
28. Ferrari, *Flora*, 192.
29. Fabio Colonna, *Annotationes et Additiones*, in *Rerum Medicarum Novae Hispaniae Thesaurus seu Plantarum Animalium Mineralium Mexicanorum Historia*. (Rome: Vitalis Mascardi, 1651), 890.
30. For a fascinating collective attempt to recreate the contents of Lastanosa's collections of art, maps, instruments, and curiosities, within the context of Huesca's religious culture, see Carmen Morte García and Carlos Garces Manau, eds. *Vicencio Juan de Lastanosa (1607-1681): La pasión del saber* (Huesca: Instituto de Estudios Altoaragoneses, 2007
31. On Lastanosa's efforts to promote the cults of St Orencio, St Prudencia, St Laurence, and St Orencio, bishop of Auch, see *Vicencio Juan Lastanosa*, 79-85; 218-221; 375-81. For another contemporary example of urban patriotism tied to the invention of relics of proto-Christian martyrs, see Katie Harris, *From Muslim to Christian Granada: Inventing a City's Past in Early Modern Spain* (Baltimore: The Johns Hopkins University Press, 2007).

CHAPTER FIVE

LOOKING AT EXOTICA IN BAROQUE COLLECTIONS: THE OBJECT, THE VIEWER, AND THE COLLECTION AS A SPACE

DANIELA BLEICHMAR

1. The elusive object

The impressive collection compiled by Vincencio Juan de Lastanosa (1607-1681) in his residence in Huesca included a wide range of objects. Going from room to room in the palace, a visitor could view all sorts of *naturalia* and *artificialia,* that is, organic and man-made objects. The collection included the types of objects considered wonders, curiosities, and rarities. Lastanosa owned antiquities such as bronzes, marbles, and coins and medals (on which he authored a treatise); a significant collection of books; mathematical and optical instruments; works of fine craftsmanship such as maps, prints, paintings, fine tapestries and decorative hangings, sculptures, and carvings; exotic objects from distant lands; and rich furnishings, from the cabinets used to house the collection to luxurious chairs, tables, and beds (a major status-signifier in early modern European homes).[1] The sense of plenty and luxury was amplified by the large number of mirrors—flat, convex, concave—that hung in various rooms reflecting and repeating the collection, turning spaces into spectacular kaleidoscopes.

We know about the items in this collection through several accounts from the mid-1600s, written in various genres. A manuscript inventory in prose produced around 1650 describes the Lastanosa residence room-by-room, as well as the garden. An exaltation in verse, with similar content, appeared in print in 1647. In the 1662 *Relación* of an encounter—real or imaginary—between Lastanosa and a visitor to the palace, Lastanosa gives the reader a virtual tour through the collection. Finally, an inventory of the library, dated to about 1640-1662, lists not only books but also many other objects that were kept in that same room.[2]

These documents attest to the presence in the collection of many objects that would fall under the category of "exotica." Lastanosa owned, among others, exquisite stones used by "Indians" as medical remedies; a shell in the shape of a boat (probably a Nautilus shell), carved with "men, birds, and plants from China;" a sculpture of a black man's head in jet (like the shell, exotic both in terms of the material and the subject); two Chinese chests with mother-of-pearl and gilt inlay; and three New World "idols."[3] It is not surprising that Lastanosa owned exotic objects, nor that he owned these specific exotica. Collectors at the time held these very objects in high regard, as their presence in inventories, catalogues, and still life painting attests. Indeed, it would have been a surprise not to find them in such a magnificent collection. However, it is far from obvious what role these objects played for Lastanosa and contemporary viewers of this or other collections. What did seventeenth-century collectors and viewers make of these artifacts?

This essay will examine some of the exotic objects in Lastanosa's and other contemporary collections as a launching pad for exploring questions about display, viewing practices, and the notion of collections as both spaces and narratives. It will also provide an opportunity to reflect on questions of methodology, examining the historical questions we can ask of objects, texts, and images. I will argue that objects in early modern collections can not be understood in isolation from one another and from the space in which collectors displayed and viewers encountered them, although this is precisely what the two textual genres associated with collections—the catalogue and the inventory—push toward. Instead, I will argue for the need of thinking of collections as spaces that constructed narratives through strategies of display and protocols that guided visitors into highly specific viewings.[4] Rather than thinking of collections as accumulations of objects, I will suggest that we view them as spaces for looking at objects.

While, as far as I know, no exotic objects from Lastanosa's collection have come down to us, we do have an image of one of them that was created at the time (Figure 1). The drawing depicts a carved ivory olifant, seen in profile. The image is accompanied by two inscriptions providing information about the object, written in different hands. At the top of the page, a legend identifies the item as "ivory horn that is in the antiquities of Vincencio Lastanosa in the year 1635."[5] Curving alongside the image, a second inscription notes, "Ivory horn measures in length a *resma* less than a *vara* it is in one piece."[6] This double inscription tells us about the artifact: what we have (a horn), the material it's made of (ivory), its size (roughly 80 cms. or 33 in.), and something about its craftsmanship (the fact that it was carved out of a single piece of ivory rather than assembled from multiple smaller pieces). More important to my discussion, the text frames the object as a collectible: who owned it, a moment in time when it formed part of the collection, and the category to which it belonged—that of antiquities and not, as we might think, exotica.

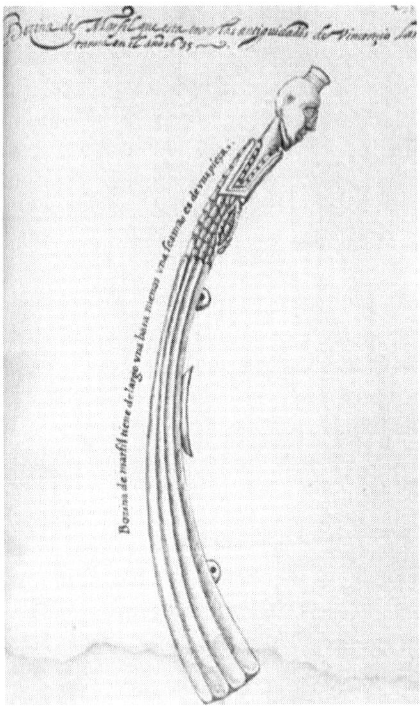

Figure 1. Drawing of an ivory oliphant from the collection of Vincencio Juan de Lastanosa. (Francisco Asín Remírez de Esparza Collection).

The issue of categories is significant for my attempt to understand how seventeenth-century viewers might have understood this object. This question turns out to be harder to answer that one might initially think, and the more information we have to inscribe the object within the collection, paradoxically, the more elusive it gets. This is how Juan Francisco Andrés de Uztarroz described this artifact around 1647, in his *Descripción del palacio y los jardines de Vincencio Juan de Lastanosa*: "Another [horn] in ivory almost a *vara* in length, from a Japanese King, two thirds striated and the last [third] covered in scales; it is crowned with a cayman head, which grabs with its mouth the head of a king."[7] All of a sudden, we have two kings, one depicted, one in Japan. The item is similarly, if more succinctly, explained in the *Relación* of 1662: "An ivory horn almost a *vara* in length from a Japanese King."[8] These two statements confirm the function, material, and size of the artifact. They also introduce a new and very interesting piece of information, one that does not have to do with the physical aspect of the artifact or with its present existence within Lastanosa's collection: the object came from Japan, where it belonged to a king. All of a sudden, what we thought was an antiquity becomes instead exotica—or rather, additionally, for the two are not mutually exclusive categories. Seventeenth-century antiquarians and collectors extended their interest in classical antiquity to a comparative evaluation of artifacts from different cultures, so that the spectrum of human cultural variability accommodated distance in time as well as distance in space. Perhaps the best known example of this type of approach is the Jesuit polymath Athanasius Kircher, with his interests in Greco-Roman, Egyptian, Chinese, and New World cultures, but other collectors shared this attitude.[9]

Additional information about the oliphant makes it even more slippery. The very same piece is itemized in an undated catalogue of Lastanosa's library, produced *ca.* 1640-1666. The entry describes: "An ivory horn almost a *vara* in length, in one piece. The [bottom] two-thirds very finely striated, in the last [third] spurts a cayman head with many scales, and from its mouth comes out the head of a crowned Indian. It belonged to an Indian king."[10] This account introduces new variations: the craftsmanship is assessed and pronounced "very fine," the human form depicted is not the head of a king but that of an Indian who is crowned (suggesting he might be a regal figure), and the king who owned the object before it entered Lastanosa's collection was not Japanese but from "India," though whether this word means the New World or Asia is unclear. This geographical instability, the very same object is considered Japanese in one statement and Indian in another, is not particular to this item, or to Lastanosa's collection. Early modern man-made and natural objects, texts, and images were often said to come from regions far from where they originated. Vague and imprecisely used terms such as "India" or "China" did not stand for specific geographic locations but for an exotic origin that was hard to pin down, remained unspecific, and could mutate unexpectedly. For instance, Lastanosa's contemporary Ferdinando Cospi (1606-1686) received in 1665 an Aztec codex as a gift for his celebrated

Bologna collection. The Mexican codex, now know as the "Codex Cospi," initially had a cover labeled "Libro della China." Another New World codex entered the collection of the Holy Roman Emperor Leopold I in 1677 or 1678. In the Hapsburg Imperial Library, the Mixtec manuscript now known as the "Vienna Codex," while recognized as coming from New Spain, formed part of a section labeled "Oriental Manuscripts."[11] Geographical imprecision was the norm rather than the exception.[12]

The ivory horn in Lastanosa's collection exhibits comparable fluidity, giving rise to multiple questions. Was it considered an antiquity? An exotic object? Or both? If so, did it inhabit the two categories simultaneously, or did it move from one to the other depending on context? The artifact's mutability extends not only to its categorization but also to its physical placement within the collection. The *Descripción* places it in the room in Lastanosa's palace dedicated to arms, while the 1662 catalogue includes it in the section on "Medals or coins and other antiquities" within the library. Finally, how would the object signify, for seventeenth-century audiences, if it were not simply a rare horn but a rare horn that had once belonged to a king, somewhere far from Europe? Would it mean something different if this king were Japanese or Indian? Geography might have been imprecise and uncertain, but gestures suggest it was not meaningless: in the sixteenth century, the Vienna Codex was said to come from "South India;" in the seventeenth century the attribution was revised to "Mexican." The inscription "Libro della China" on the Codex Cospi was revised to "Libro del Messico," and it is as such that the item is described in the published catalogue (1677) and inventory (1680) of the collection.[13] Origin was important enough to merit debate and revision.

For the historian, the issue is not one of precision, of whether they "got it right." My intention is not to bring into question the veracity of early modern descriptions, even less to pin down the identity of a particular object—to figure out exactly where the horn was from (even though we can be fairly certain it is of African origin, based on stylistic criteria). Rather, I want to make both a methodological point about the ways in which we read early modern accounts of collections, and an interpretive one, by suggesting that the slipperiness of early modern collectibles originated not in their descriptions but in the objects themselves, or rather, in the possible responses available to viewers. If we have inconsistent accounts of objects like Lastanosa's ivory horn, it is precisely because they could be understood and described in multiple ways. I should make clear that I am not referring to the ways in which objects in early modern cabinets constantly blurred the line between the natural and the artificial, the scientific and the artistic, the antique and the exotic.[14] This hybridity and categorical fluidity can be exemplified, for instance, by the rhinoceros cups highly so prized by European collectors in the sixteenth and seventeenth centuries. The material out of which these objects were crafted, rhinoceros horn, made the cups both *naturalia* and rarities; they were also objects of wonder, since rhinoceros horn purportedly had magical properties, as well as exotic ones, since the material came from outside Europe, most probably from Goa after the Portuguese conquest of 1540. In addition, rhinoceros cups often included elaborate gilt silver mounts and lids—made of a material both natural and precious but one crafted by man. These mounts would often refer to the natural through floral or zoological decorative patterns, and also served as evidence of *artificialia* and of fine craftsmanship.

But, going beyond the hybridity inherent in the objects, I want to point to something else: not the multiple categories in which an object could fit, but rather the multiple ways in which the same types of objects, or even a single object, could be interpreted. I would argue that early modern collectibles were fundamentally multivalent and variable not only in their corporality but also in their interpretability.[15] Artifacts would have yielded various interpretations and associations according to the specific settings in which they were embedded, that is, the contexts in which they were displayed, whether religious, courtly, civic, professional, institutional, and so on. Early modern collections show remarkable consistency in the types of objects they included—the list is fairly standardized, something attributable to both exchange networks (whether of trade or gift-giving) and the referentiality created by print culture. However, the possible interpretations of objects were strongly context-dependent. A feather cape from the New World could signify according to multiple political, artistic, historical, religious, sociological factors depending on who owned it, what that person made of it in his or her collection, and who viewed it. These symbols were overdetermined, making automatic or transparent iconographic readings impossible.[16] According to Martin Kemp, "much of the life and sustained fascination of the collections for highly sophisticated viewers lay in the refusal of many of the individual objects to submit docilely to precise categorization."[17] The issue, then, is to explore what seventeenth-century collectors, viewers, or readers would have made of particular objects within the specific contexts in which they encountered them. The history of collecting, collections, and collectors needs to be, to some degree, a history of display and of reception.

Questions of meaning and interpretation have been central to the history of collecting. Paula Findlen has described sixteenth- and seventeenth-century Italian natural history collections as spaces for the production of knowledge and the construction of social networks. Thomas DaCosta Kaufmann has analysed the spectacular collections of Holy Roman Emperor Rudolph II in Prague according to the notion of *representatio*, which implied a symbolic mastery of the world connected to political power. Kryzstoff Pomian has studied the changes in the cultural and social history of collecting and the persona of the collector in France and Italy between 1500 to 1800. Joy Kenseth has framed sixteenth- and seventeenth-century cabinets within a culture of curiosity and wonder, a topic richly expanded by Lorraine Daston and Katharine Park's magnificent study on wonder and the order of nature.[18] But, in general, the scholarship on early modern cabinets has paid scant attention to issues of display and viewership— to the question of looking, and of the type of looking that the collection as a space required.[19]

2. The viewer in the cabinet

Trying to trace a history of displaying and seeing collections in early modern Europe is far from easy. Viewers turn out to be no less elusive than objects. For all their fascination with cabinets, early modern writers left precious little testimony of what collections actually looked like or how viewers perceived them. Collectors emphasized the interest their cabinets arose, but mentioned almost nothing about display or viewers' responses. For instance, Francesco Imperato gives us barely anything to go on when he wrote of his father's famed collection, "There is not a Signor or great person who comes to Naples from far and remote parts," he reported, "who does not wish to see [the museum] out of curiosity and who is not overcome with amazement, after having seen it."[20] Accounts from viewers are rare; those we have tend to be similarly vague. A visitor to the collection of Ole Worm (1588-1655) in Denmark stated in 1654 that in it, "is found and can be examined with wonder, odd and curious rarities and things among which a large part has not been seen before, and many royal persons and envoys visiting Copenhagen ask to see the museum on account of its great fame and what it relates from foreign lands, and they wonder and marvel at what they see."[21] As with Imperato, we know nothing beyond a register of characters of notable standing and the tropes of curiosity, rarity, marvel, and wonder. There is no mention about display or about the collection as a space—even that is what might be most striking to a modern viewer of the prints depicting these cabinets (Figure 2).

Figure 2. Francesco Imperato, *Historia naturale di Ferrante Imperato napolitano; nella quale ordinatamente si tratta della diuersa condition di minere, pietre pretiose, & altre curiosità; con varie historie di piante, & animali, fin'hora non date in luce* (Venice, 1672). Research Library, The Getty Research Institute, Los Angeles, California.

We get more insight from Thomas Platter, a native of Basel who wrote about seeing Sir Walter Cope's renowned collection of curiosities in London in 1599. Platter's account of the visit starts: "This same Mr. Cope inhabits a fine house in the Snecgas; he led us into an apartment, stuffed with queer foreign objects in every corner, and amongst other things I saw there, the following seemed of interest." [22] Several things are noteworthy about this statement: that the collection occupied a discrete space within Cope's residence, constituting a cabinet as such (rather than spilling out throughout the entire residence, as was the case with Lastanosa's palace); that the collector personally took visitors into this space and lead them through his collection (as Lastanosa's *Relación* also suggests); that Platter immediately remarks on rarity and exotic origin as the significant attributes of the objects, and (at long last!), something on display: the space is crowded, the aesthetic that prevails is one of plenty and accumulation (there are "queer foreign objects in every corner"). Interestingly, at the very moment in the travel narrative in which Platter enters the cabinet, as soon as he crosses that door, he immediately abandons prose and switches to an inventory format, as if this were the genre best suited to writing about collections. He tells us, "Amongst other things I saw there, the following seemed of interest," and then proceeds to list fifty numbered items. "1. An African charm made of teeth. 2. Many weapons, arrows and other things made of fishbone. 3. Beautiful Indian plumes, ornaments and clothes from China" (note India and China are uttered in the same breath). He goes on to mention such notable items as "9. Beautiful coats from Arabia… 12. The horn and tail of a rhinoceros, is a large animal like an elephant… 16. A round horn which had grown on an English woman's forehead… 19. The bauble and bells of Henry VIII's fool; 20. A unicorn's tail [not its horn]… 33. Porcelain from China; 34. A Madonna made of Indian feathers; 41. A mirror which both reflects and multiplies objects;" ending with "50. A long narrow Indian canoe, with the oars and sliding planks, hung from the ceiling of this room." After this inventory, Platter returns to prose with a speedy wrap-up, noting that Cope also, "possesses besides many old heathen coins, fine pictures, all kinds of corals and sea-plants in abundance."[23]

The list focuses on those items that Platter found most noteworthy. This process of selection does not give any indication of the total scope of the cabinet—we do not know if Platter lists 50 items out of a total of 60, 300, or thousands. It is clear that Platter as a viewer was interested in the curious rather than on antiquities, paintings, or naturalia; he dispatches the latter rather quickly at the end of the list, not showing much interest. The objects he itemizes evidence the characteristic attributes of the culture of curiosity: rarity, uniqueness, geographic origin, provenance from a famous person, beauty, medium, material, craftsmanship, and scale. Themes reappear throughout the list, mostly in terms of shape or material: feathers, teeth, stones. Objects across the collection resonate with one another, visual motifs are repeated as if reflected in multiple mirrors: there is a horn from a rhinoceros, another from a bull seal, yet another from a woman's forehead.

However, Platter does not detail how items were stored or displayed, nor does he indicate what was placed next to what, whether arrangement was based on typology, size, material, some other principle, or a combination of factors.[24] We do not know the relationship between the order in which items appear in the list and their spatial relation within the cabinet. Did Platter number entries in the order he saw objects, moving down the list as his gaze moved from one part of the room to another?

Though it's impossible to know, it seems highly unlikely that he would have been putting pen to paper while Cope led him in a visit through the cabinet, and if we assume he wrote the list post-facto then he most probably would have been deploying strategies based on the theater of memory tradition—that is, putting into practice the idea of the cabinet as both a theater of the world and a theater of memory as expressed in Samuel Quiccheberg's 1565 treatise on collecting, which in turn drew heavily on Giulio Camilo's work on memory.[25] This would imply that there is at least some relation between the order in which items appear in the list and the way in which objects were displayed in the cabinet.

Platter provides only two pieces of information about display: item 44 lists, "Saddles from many strange lands; they were placed round the top on stands." This suggests that there was some sort of cabinetry or shelving, perhaps high, so that stands could be placed on top of it. It also reveals that certain items were not contained within specific sections of the cabinet but spread around it. This is reminiscent of what we see in the depiction of Cospi's cabinet (Figure 3), in which some sculptures were placed on top of the shelving, and arms were spread out across the room in this top-most space. The same exhibitionary principle applies to medals, which form a distinct bottom layer around the room. Item 50 in Platter's list, the "indian canoe," hung from the ceiling—not unlike the kayak that hung from the ceiling in the cabinet of Ole Worm (Figure 4). I should make clear that these depictions of Cospi's and Worm's collection cannot be taken at face value as literal renditions of what their cabinets actually looked like. As all prints of early modern cabinets, they must be understood as an idealized rendition of a collection in a state in which it most probably never existed. However, they do provide a sense of the display aesthetic that was also at work in Cope's and Lastanosa's cabinets, and of practices of looking that resonate with Platter's list.

Just like contemporary practices for displaying and viewing art are interconnected, with protocols of spatial arrangement for both art objects and their visitors, and specific practices of seeing, early modern spaces called for specific viewing practices. Platter's list conveys a particular type of viewing, one that is also in evidence in the texts documenting Lastanosa's collection and in seventeenth-century images of cabinets. This is a way of seeing in which the eye is in constant motion from object to object, zooming in and out. A visitor might first look around the room to get a sense of the cabinet as a whole, as a space or as a composite artwork that brings together multiple parts *(Gesamtkunstwerk)*. The visitor would next focus on a specific portion of the collection—the animals hanging from the ceiling, or the group of small items in one box or shelf—only to

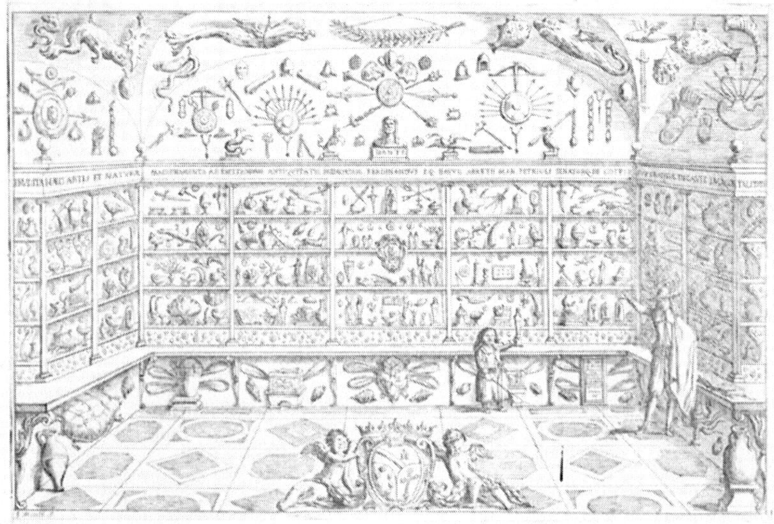

Figure 3. Lorenzo Legati, *Museo Cospiano: Annesso a Quello del Famoso Vlisse Aldrovandi e Donato alla sua Patria dall'Illustrissimo Signor Ferdinando Cospi* (Bologna, 1677). Research Library, The Getty Research Institute, Los Angeles, California.

then shift his attention to a single object, perhaps a medal, a shell, or a stone that can be held in one's hand, brought towards the body for close inspection. But the eye would soon sprint back to another object, near or far in space, that resonates with the first one; and so on. It's an agile and darting eye, and one that constantly shifts both focus and speed from wide to narrow, from slow to fast.

Like Platter's list, Uztarroz's *Descripción* suggests a display strategy based on plenty, with "queer foreign objects in every corner." But Lastanosa's collection far exceeded Cope's compact cabinet. His belongings did not occupy a single room but spilled over the entire palace in a cornucopia of luxury that extended from room to room, from wall to wall, from floor to ceiling. The aesthetic is a baroque one of ornamentation and *horror vacui*, and though no representations of Lastanosa's collection exist it is clear that it had much in common with contemporary cabinets such as those of Francesco Calceolari or Manfredo Settala, as evidenced in the prints depicting these two Italian collections (Figures 5 and 6).[26] The aesthetic is maximalist, more-is-more. The arrangement is strongly

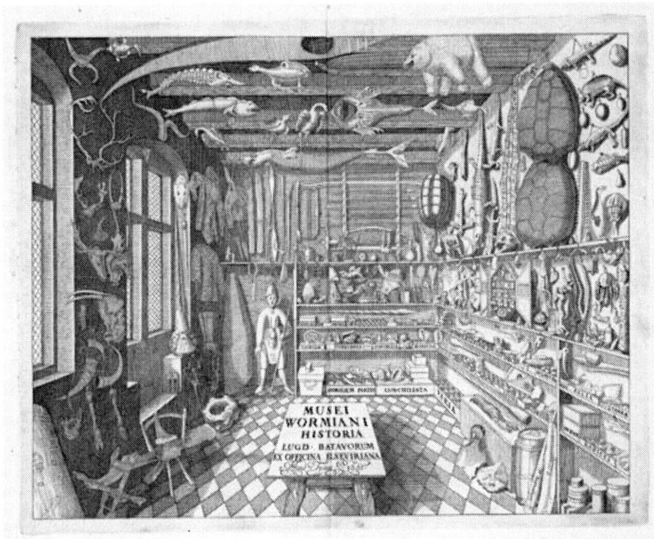

Figure 4. Ole Worm, *Museum Wormianum, seu, Historia rerum rariorum: tam naturalium, quam artificialium, tam domesticarum, quam exoticarum, quae Hafniae Danorum in aedibus authoris servantur / adornata ab Olao Worm...; variis & accuratis iconibus illustrata* (Leyden, 1655). Research Library, The Getty Research Institute, Los Angeles, California.

symmetrical, and there is a delight in the richness of materials—marble floors, finely and richly crafted shelves (in Calceolari) or smaller cabinets (in Settala). There is a predilection for variety by alternating shapes, media, types of objects. We can also see a hint of groupings, of collections within the collection: Calceolari's line-up of birds, perched atop the shelves that run along either side of the room, or the classical structure at the end of the room, with niches to hold small figurines; Settala's own line-up of paintings (most probably, though they could be wall-hangings) and of the cabinets that hold items). There is so much information that the viewer needs to look bit by bit in order to see the birds that appear to be about to fight with the animals, the furious movement produced by the creatures hanging from the ceiling (one can almost hear them!); the way in which shapes and materials echo, with the curving scroll motifs that decorate the sides of the furnishing suggesting the shapes of the shells on a shelf, or the fluted ridges of a shell repeating the shape of an urn or the spines of books; the fact that drawers burst open to reveal neatly lined up series of minimally varying objects or, somewhat terrifyingly, the neatly ordered rows of cabinet after cabinet, each of which would open up, like a Russian doll, to reveal yet more collections, so that what appeared like excess turns out to be restraint, and the unrestricted outpouring of so many things onto the page, which seemed to imply that nothing was being held back, actually reminds the viewer that there is still more and more. Despite its plethora, the baroque cabinet does not show everything; it gradually reveals through careful looking.

With this in mind, we can return to Lastanosa's ivory horn (Figure 1). The blank page that isolates the item, like an island floating on all that white space, now looks incongruous. How can we reinscribe the horn within a display space in which viewers would have encountered it, and a type of looking? We have a hint in its entry in the *Descripción:* the first word is "another." The horn was never meant to stand alone but was part of a larger group of objects, and would have been seen in relation to them. Uztarroz mentions the horn within his description of the room in Lastanosa's palace that he calls the armory. He writes:

Figure 5. Benedetto Ceruti, *Musaeum Francisci Calceolarii* (Verona, 1622). Research Library, The Getty Research Institute, Los Angeles, California.

Figure 6. Paolo Maria Terzago, *Museo, ò Galeria, adunata dal sapere e dallo studio del sig. canonico Manfredo Settala nobile milanese / descritta in latino dal sig. dott. fis. coll. Paolo Maria Terzago; et hora in Italiano dal sig. Pietro Francesco Scarabelli* (Tortona, 1666). Research Library, The Getty Research Institute, Los Angeles, California.

There are in this same room three shelves with harquebuses and muskets with all their trimmings, some of them rare due to the shape of the barrels and boxes, and a variety of exquisitely manufactured pistols; two shelves of pikes, blow darts for talking at a distance; on an entire mat is hung and dressed with arms of great antiquity, war instruments, hunting crossbows, quivers with madrazos, war crossbows with quivers with arrows and darts with little lances, and steel heads. Hunting and fighting bows, the former with turquoise bodoques, the [latter] war ones with quivers and arrows. Ivory [shooting] rings. A bronze horn. **Another [horn] in ivory almost a vara in length, from a Japanese King, two thirds striated and the last [third] covered in scales; it is crowned with a caiman head, which grabs with its mouth the head of a king.** An estorica [crossbow?], weapon of the Queen of the Amazons, mentioned by Father Acuña in the memoir already mentioned. Two sabers, one from the King of the Tatars, garnished with tortoiseshell and gilded bronze, with most exquisite relief work; given to don Vincencio Juan de Lastanosa by his Excellency don Fernando de Gurrea Aragon y Borja, Duke of Villa Hermosa. The other saber is from China, the blade like a ray, entirely engraved with waves, it ends with a little devil used as the hilt.

A dagger of King Don Pedro the III of Aragon [1239-1285], with which he cut his fingers when he cut the Privileges of Union. There is a copious variety of most antique arms, such as bibs, breastplates, a steel shield from Milan, another made of wood from Salamanca, many other items such as are used in war as well as in jousting and tournaments...[27]

Thus, even if catalogues and inventories present us with individual objects, we must think of the collection as a whole, as a space, and as an experience. The gaze operative in the collection, traveling from object to object, was a way of seeing that constructed narratives. The collector who guided a visitor through his cabinet, doubtlessly told him or her stories about the objects on display. Many images represent the collection not only as an accumulation of objects but also as a narrated social experience (see Figures 2 and 3). The guide whom we see in seventeenth-century prints of cabinets, often gesturing and wielding a pointing stick, invites us not so much to see objects as to see them in a certain way and as part of a certain story. Visitors to baroque collections experienced them as narratives, constructed by the collector and by themselves. It was these narratives that pinned down, even if fleetingly, the meanings of objects.

Notes

1. Beds were often the most expensive possession listed in household inventories; luxurious beds such as are described in Lastanosa's residence could be worth more than works or art or other collectibles. See Rafaella Sarti, *Europe at Home: Family and Material Culture, 1500-1800* (New Haven: Yale University Press, 2002) and Patricia Fortini Brown, *Private Lives in Renaissance Venice: Art, Architecture, and the Family* (New Haven: Yale University Press, 2004).
2. Juan Francisco Andrés de Uztarroz, *Descripción del palacio y los jardines de Vincencio Juan de Lastanosa* [1650?], Hispanic Society of America (New York), MS. B-2424, f. 24r.-51v; Juan Francisco Andrés de Uztarroz, *Descripción de las Antigüedades i Jardines de don Vincencio Iuan de Lastanosa, hijo i ciudadano de Huesca, ciudad en el Reino de Aragon* (Zaragoza, Diego Dormer, 1647); *Narración de lo que le pasó a Don Vincencio Juan de Lastanosa a 15 de octubre del año 1662 con un religioso docto y grave*, Hispanic Society of America (New York), MS. B-2424, f. 52r.-79v; *Catálogo de la biblioteca de Vincencio Juan de Lastanosa* (Sparvenfeld Catalogue [1640-1662?]), MS. U-379, Royal Library, Stockholm. I am extremely grateful to the organizers of the conference that yielded this volume for making their transcriptions of these four documents available to me. All mention and citations of these documents in my paper refer to these transcriptions, not to the originals. I have ignored a fifth account of the collection, *Las tres cosas más singulares que tiene la Casa de Lastanosa en este año de 1639,* Biblioteca Nacional de España, MS. 18727-45, since the organizers' have questioned its authenticity. All translations in this essay are mine.
3. "mui esquisitas, de que se valen los indios para el remedio de sus enfermedades," Uztarroz, *Descripción del palacio y los jardines,* f. 28r; "hombres, aves y plantas de la China," 26r.; "cofrecillos de casi de una bara de obra de la China, el campo negro con laços de nacar o madre perla, adornados de cazas y paises de oro molido, cerrajas y aldavas doradas," 37v.

4. The idea of the collection as a narrative is explored in Mieke Bal, "Telling Objects: A Narrative Perspective on Collecting", in John Elsner and Roger Cardinal, *The Cultures of Collecting* (London: Reaktion Books, 1994), pp. 87-116.
5. "Bozina de Marfil que esta en las antiguidades de Vincençio Lastanosa en el año 1635."
6. "Bozina de marfil tiene de largo una bara menos una resma es de una pieça."
7. "Otra [bocina] de marfil de casi una bara de largo, de un Rey del Japon, los dos tercios estriados y el ultimo escamado; remata en una cabeza de cayman, tiene asida con la boca la cabeza de un rey," Uztarroz, *Descripción del palacio y los jardines,* f. 47r.-47v.
8. "Una bocina de marfil de casi una bara de larga de un Rey del Japon," *Narración de lo que le pasó a Don Vincencio Juan de Lastanosa*, f. 75v.
9. Among the vast literature on Kircher, see Paula Findlen (ed.), *Athanasius Kircher: The Last Man Who Knew Everything* (New York: Routledge, 2004) and Daniel Stolzenberg (ed.), *The Great Art of Knowing: The Baroque Encyclopedia of Athanasius Kircher* (Stanford, CA: Stanford University Press, 2001).
10. "Una bocina de marfil de casi una bara de largo de una pieça, los dos tercios estriada con mucha gala, en el ultimo hecha una cabeza de cayman con muchas escamas, y de la boca le sale una cabeza de un indio coronado, era de un Rey de la Yndia." *Catálogo de la biblioteca de Vincencio Juan de Lastanosa*, Royal Library, Stockholm, MS. U-379, f. 102v., consulted in Garcés Manau transcription. Dating, Mar Rey Bueno, personal communication.
11. I thank Arianna Opsvig for bringing these two examples to my attention.
12. Benjamin Schmidt discusses the exotic mode in the context of Dutch visual culture at the time in, "Inventing Exoticism: The Project of Dutch Geography and the Marketing of the World, circa 1700," pp. 347-369. For an illuminating discussion of the role of geography in the history of art, see Thomas DaCosta Kaufmann, *Toward a Geography of Art* (Chicago: University of Chicago Press, 2004).
13. Lorenzo Legati, *Museo Cospiano: Annesso a Quello del Famoso Vlisse Aldrovandi e Donato alla sua Patria dall'Illustrissimo Signor Ferdinando Cospi* (Bologna, 1677), book 3, chapter 2, p. 191, and *Inuentario semplice di tutte le materie esattamente descritte che si trouano nel museo Cospiano: non solo le notate nel libro gia stampato e composto dal sig. Lorenzo Legati, ma ancora le aggiunteui in copia dopo la fabrica* (Bologna, 1680).
14. See for instance, Horst Bredekamp, *The Lure of Antiquity and the Cult of the Machine* (Princeton, NJ: Markus Wiener Publishers, 1995).
15. My thinking on this point owes much to Martin Kemp, "'Wrought by No Artist's Hand': The Natural, the Artificial, the Exotic, and the Scientific in Some Artifacts from the Renaissance," in Claire Farago (ed.), *Reframing the Renaissance: Visual Culture in Europe and Latin America 1450-1650* (New Haven: Yale University Press, 1995), pp. 175-196.
16. Kemp argues that, "a complex fluidity, ambiguity, and diversity of meaning characterizes the viewing of such items even in a number of apparently similar contexts in Renaissance societies, and that such viewing undermines any propensity to characterize them neatly in terms of the kind of historical "meta-

realities"—such as power, colonialism possession, oppression, patriarchy, Eurocentrism, and otherness—which now tend to be taken as having a privileged explanatory power," in "'Wrought by No Artist's Hand'," pp. 180-1.
17. Kemp, "'Wrought by No Artist's Hand'," p. 185.
18. Paula Findlen, *Possessing Nature*, (Berkeley, Los Angeles and London, 1994); Thomas DaCosta Kaufmann, *The mastery of nature: aspects of art, science, and humanism in the Renaissance* (Princeton, N.J.: Princeton University Press, 1993), esp. ch. 7 and *Court, Cloister, and City: The Art and Culture of Central Europe, 1450-1800* (Chicago: University of Chicago Press, 1995); Kryztoff Pomian, *Collectors and Curiosities*: Paris and Venice, 1500-1800 (Cambridge, UK: Polity Press, 1990); Joy Kenseth (ed.), *The Age of the Marvelous* (Hanover, NH: Dartmouth College Art Museum, 1991); Lorraine Daston and Katharine Park, *Wonders and the Order of Nature,* 1150-1750 (New York: Zone Books, 1998).
19. This is in stark contrast, it bears mentioning, with the literature on contemporary museum studies, which tends to focus on exhibition techniques and visitor's experiences. See for instance Ivan Karp and Steven D. Lavine, *Exhibiting Cultures. The Poetics and Politics of Museum Display* (Washington and London, 1991), esp. Michael Baxandall, "Exhibiting Intention: Some Preconditions of the Visual Display of Culturally Purposeful Objects." The classic discussion of the importance of situating the eye historically is Michael Baxandall's powerful concept of the "period eye," in *Painting and Experience in Fifteenth Century Italy: A Primer in the Social History of Pictorial Style* (Oxford and New York: Oxford University Press, 1988 [1972]). On space-specific viewings in early modern Europe, see for example Randolph Starn, "Seeing Culture in a Room for an Renaissance Prince," in Lynn Hunt (ed.), *The New Cultural History* (Berkeley: University of California Press, 1989), pp. 205-232.
20. Francesco Imperato, *Discorsi intorno a diverse cose naturali* (Naples, 1628), sig. 3, quoted in Findlen, *Possessing Nature,* p. 31.
21. Joy Kenseth, "A World of Wonders in One Closet Shut," in Joy Kenseth (ed.), *The Age of the Marvelous* (Hanover, NH: Dartmouth College Art Museum, 1991), pp. 80-101, quoted on p. 81.
22. Reproduced in Peter C. Mancall (ed.), *Travel Narratives from the Age of Discovery: An Anthology* (New York: Oxford University Press, 2006), pp. 401-403.
23. *Ibid.*
24. Objects in large princely collections were often arranged according to their material—this method was recommended by Quiccheberg, and practiced in the Munich cabinet of Duke Albrecht V (1528-79), and Ferdinand II's at Ambras Castle. Size and shape were also important governing principles. See Kenseth, "A World of Wonders in One Closet Shut," pp. 84-85.
25. Samuel Quiccheberg, *Inscriptiones, vel, Tituli theatri amplissimi, complectentis rerum universitatis singularas materias et imagines eximias* (Munich, 1565). On the memory theater, see Frances Yates, *The Art of Memory* (Chicago: University of Chicago Press, 1966).
26. Ceruti, Benedetto, *Musaeum Francisci Calceolarii* (Verona, 1622) and Terzago, Paolo Maria, *Musaeum Septalianum* (Tortona, 1664).

28. "Hay en esta misma pieza tres estantes de arcabuces y mosquetes con todos sus aderezos, algunos raros por la forma de los caños y cajas, y variedad de pistolas de exquisitas hechuras; dos estantes de alabardas, cerbatanas para hablar de lejos; en toda un estera se ve colgada y vestida de armas de mucha antigüedad, instrumentos belicos, ballestas de caça, carcajes con madrazos, ballestas de guerra con carcajes con flechas y dardos con lancillas, y casquillos de acero. Arcos de caça y pelea, los primeros con turquesas para hacer los bodoques, los de guerra con aljavas y flechas. Anillos de marfil. Una bocina de bronce. Otra de marfil de casi una bara de largo, de un Rey del Japon, los dos tercios estriados y el ultimo escamado; remata en una cabeza de cayman, tiene asida con la boca la cabeza de un rey. Una estorica, arma de la Reyna de las Amazonas, de la qual hace memoria el Padre Acuña en la relacion referida. Dos catanas, una del Rey de los Tartaros, guarnecida de carai y bronce dorado, con primorosissimas lavores de relieve; diosela a don Vincencio Juan de Lastanosa el Excelentisimo Señor don Fernando de Gurrea Aragon y Borja, Duque de Villa Hermosa. La otra catana es de los chinas, la cuchilla como rayo, toda gravada de aguas, remata esta y tiene y sirve de empuñadura un diablillo. / Un puñal del Rey don Pedro el IIII de Aragon, con que se cortó los dedos quando cortó los Privilegios de la Union. Ay otra copiosa variedad de armas antiquissimas, como son petos, coraças, rodela de acero de Milan, de madera de Salamanca, otras muchas pieças assi de las que se platican en la guerra como en las justas y torneos…," Uztarroz, *Descripción del palacio y los jardines,* f. 47r.-47v. My emphasis.

CHAPTER SIX

FRIENDS, SCHOLARS, COLLECTORS: THE SECOND CIRCLE OF LASTANOSA

MIGUEL LÓPEZ-PÉREZ

The "circle of Lastanosa", as historians calls it, was diverse and constituted by artists, such as Ana Abarca of Bolea (1602-1685), Jusepe Martínez (1600-1682), Francisco de Artiga (1645-1711), Manuel Salinas (1615-1688) and Baltasar Gracián (1601-1658); collectors and scholars, such as the Count of Guimerá (1584-1638); and professional jurists and, historians, such as Diego Vincencio Vidania (1650?-1711), Juan Francisco Montemayor (1620-1685), Diego Jose Dormer (16??-1703), and Andrés de Uztarroz (1572-1635). If we understand as "circle" the group of people who habitually shaped the set of relations they supported with each other along their lives, either simultaneously or not, then, the people we have just mentioned constituted the Circle of Vincencio Juan of Lastanosa. But, if we pay attention to whom he played the central role in the circle and its activities, such as collecting with pleasure, the interest for the garden, the creation of a library and the search for fabulous medicines for alchemy, certainly, we will be able to say that the role that played by the "circle of Lastanosa" in those activities was small indeed. In ways we are still unable to explain, Lastanosa managed to collect in his house at Huesca the most important books of the matters he was interested, and the objects most sought after among collectors of his time; he also managed to create a fashionable garden, and an alchemical laboratory. In order to create the vast collection of objects, books, and medicines and to create a library, a garden, and a laboratory it is necessary to have financial and human resources. Lastanosa had the financial resources, as well as social connections that allowed him to achieve his aspirations. We will attend to the second aspect, to those who, direct or indirectly, collaborated in the unfolding of Lastanosa's activities and occupations. Then, we will be able to glimpse another "circle", another group of persons who also developed more than one activity similar to the “Lastanosa circle”.

One of Lastanosa’s friends who helped him gather his collection was François Filhol (1583?-1648)[1]. After taken root in Toulouse, Filhol began a very particular relation with Lastanosa. Separated by not more than four hundred kilometres, this proximity was sufficient for both to establish a relationship; they were joined by something more than an interchange of the description of their collections[2]. They were pleased, directly or indirectly[3], with each other? At some point, Lastanosa received a present from Filhol, a book[4], sent from Toulouse, exactly, from the cloister of the Cathedral of Saint Mersin[5], where Filhol had deposited his collections and garden[6]. But, beyond the friendship, or of the mutual acknowledgements of their collections and gifts, what is of greater interest is the contribution that Filhol made to the garden of Lastanosa, especially when Filhol sent him seeds of tulips. This so concrete fact reveals to us a few reticulated relations, a few interconnections very interesting for the historian:

> Los Tulipanes que la Francia cria, / i tu curiosidad prodiga embia, / desarrollan aqui vistosamente / tanto esplendor luciente, / i tanta variedad, que no ai colores / que puedan dibuxar sus resplandores.[7]

The seeds of tulips sent by Filhol to Lastanosa enriched the beauty of Lastanosa’s garden[8]. The importance of the tulip in the European gardens has been successfully analyzed by some historians, and unveils to us a few consequences beyond the decorative culture around these flowers[9]. Filhol also sent Lastanosa his description of his collection——a catalogue of his collection. it is necessary to clarify that Filhol exchange with Lastanosa does not imply a special interest from Filhol to Lastanosa, or vice versa. The description of Filhol collection was done by Filhol himself for the Duke of Orleans. And the above mentioned collection provoked the interest of more persons, like Pierre Borel (1620-1689), who described it in the year 1649[10]. And also, we have to say that, unlike Huesca case, in Toulouse they were about fifteen collectors in these years,[11] which generated a strong interest in Toulouse’s collections. It is not surprising that, in the same way that Lastanosa and Filhol themselves, Borel was a collector, and he was a great Gallic scholar interested in botany, medicine and alchemy, managing to publish nothing less than the famous Bibliotheca Chimica[12].

Lastanosa’s connections do not stop here. Lastanosa himself could have had relations with Gastón de Orleáns (1608-1660)[13], who managed to have one of the best collections of his time[14]. Lastanosa also followed word-for-word what other experts of his time said on the flowers and plants growing in his garden?, and on their own gardens. Moreover, Lastanosa kept in touch with some of them, who sent him books, seeds, and exquisite and very rare flowers[15], such as Jean Baptista de Dru[16] and Pierre Morin[17], two recognized gardeners[18] and collectors[19].

All this data gives us precise and necessary news about the relationship among prominent figures similar to Lastanosa and its morphology. In this way, we can see that relationships among collectors were characterized by the existence of several elements. First, these relationships needed an issuer of the knowledge and a receptor, a receiver; that is to say, those relationships were based on communication. Second, these relationships required the presence of knowledge itself, and by any of its ways or warehouses. And finally, these relationships were shaped by the desire and the availability of the members to circulate. The depository of this knowledge was, initially, the issuer. But the recipient, simultaneously that practises as such, could also act as diffusive focus of the knowledge received. The accumulation of knowledge was not anything static; on the contrary, it was dynamically and actively increased. Neither was this knowledge closed nor inaccessible, but rather accessible and opened to the people in the relationship in which this knowledge circulated. But only this knowledge became open and accessible after a selective and more or less demanding process, shaped by multiple variables, such as economic or religious. The knowledge could be directly originated by the issuer himself, turning him into primary source, or could be indirectly originated, for example, sending information obtained from a third party. Primary source, for instance, was Pierre Morin or Jean Baptista de Dry sending books; and secondary source, for instance, was Uztarroz, who reported to Filhol as about Lastanosa's collection. And the unidirectional flow was represented by Morin sending seeds to Lastanosa, to whom, as far as we know, did not even sent anything to? Dru, as happened with Kircher, as we will see later on. Another of Lastanosa unidirectional sending of these extravagancies of the Nature was Vincenzo Mariscotti, a poetry fan[20]:

> En Bolonia con el Illustrissimo Sr. Conde Vincencio Mariscoti, que aviendole pidido la rosa senente o de la China me ofrece remitirla, aunque es tan singular que solo se halla en manos de dos Principes en Italia, y me remite una oja de la planta y ofrece que vendra acompañada de la parra o vid de la China, por ser de lo mas raro que ay en la naturaleza.[21]

The desire of the parts to circulate the above mentioned knowledge among the chosen ones stays inside a spirit of comprehension, cooperation and solidarity, mutual or not. What is what was requested? What was interchanged? What was offered? The answer to these questions depends on the type of knowledge and on the area in which someone was interested. So we find several answers to this question. In any case, is common the respect and the pretension of efficiency. They are like two natural requirements. The veracity, the validity comes at the moment of offering, since not only knowledge that was interchanged: this knowledge was currently accompanied of experiences and resources, which were seen as a set, a patrimony to corresponding effectively.

In the seventeenth century one was consciously of the power that the printed book had in its commercial, intellectual, and political and individual aspects[22]. The accumulation of books in a particular bookshop during the early modern period depended on both the content of the book and its the utility[23]. It is clearly that the interests of people like Lastanosa provoked the existence of specific books. The wish to collect exotic rarities, to establish a garden, to collect objects of the classic Antiquity, and to collect medical remedies in vogue helped to generate a printed corpus of proper, and simultaneously double, typology. Printed catalogues where these collectors saw their belongings reflected in detail, began to appear in the market. These catalogues were produced in two ways. One, by the collectors, because they were entrusting editing the catalogues themselves, or because the collections were acquiring so many reputation that others were deciding to do it, either in a explicit wish, as Manfredo Settala's collection described by Paolo Maria Terzago (d. 1695)[24], or what was doing Bernard de Montfaucon, (1655-1741) after travelling for Italy[25]. At the same time printed books describing gardens and manual of agriculture appeared in the 17th century[26]. Indeed, Lastanosa supported contacts with Manfredo Settala (1600-1680), son of the famous physician Ludovico Settala (1552-1633)[27] and the second personage on whom we are going to centre, according to what Diego Vincencio Vidania tells us[28]. This collector of Milan[29], considered "the Arquimedes of his time" by Filippo Picinelli (1604-1678)[30], managed to assemble one of the best collections of curiosities of 17th century. Sometimes he was considered as a magician and an alchemist wrongly in some occasions. But he reached along his life an enormous labour capable of provoking the desire of its continuity and remaining assembled. At least, this s what we can read in his testament, dated on July 30, 1672, where he leaves as executor to the friar Carlos. Settala orderer to Carlos in case of the disappearance of his family, that all his Museum was going on to the Ambrosiana Library[31]. For twenty years, from 1640 to 1660, his patrimony was inventoried and drawn all objects, divided in naturalia, artificialia and mirabilia, to complete three haughty volumes, of an incalculable scientific and historical value. Drawings are really impressive and they were realized by, at least, a dozen of artists' expert hands of which we know seven names: Domenico Tencala, Francesco Porro, Giovanni Battista and Francesco Volpino, Alfonso Costa, Carolo ' to Sole ' and Carolo Galluzio, being the rest only signed. Two of the books remain in the Estense Library at Modena, and the third one in the Ambrosiana Library of Milan[32].

Both in the library and in the bookshop, or art gallery, there were contributions realized by other prominent figures, whose relation with Lastanosa is not well known by us. But, we will be able to deduce, at least, some important details. Certainly, we prune to discard, for example, the accomplishment of meetings with teachers and university professors, and with colleagues for interchange of theoretical and practical knowledge. The study of some painters' names and the presence of certain pictures is also a interesting point. Out of the sponsorship of the Church, the increase of private patrons and the zeal of collecting incited the painters to make a production of small or medium format to increase the cabinets of curiosities of rich merchants, and local nobility. In this fever directly enters the care of the image, which practical effects we see in busts and portraits[33]. We must remember that nobility, during the 17th century, considered to be very important the diffusion of the ideal image, a full model of specific connotations that had to adorn the nobleman, such as consideration with the truths of the catholic faith[34], the reputation as corollary to a virtuous life, or the artistic patronage exercised as a prestigious way of life[35]. It is the time when the Art is collected in the same way as are hoarded scientific objects or the naturalia that were coming from West Indies and America. This way, Lastanosa had painters' pictures from those with scanty fortune, but not fewer artful, who never were precipitated to the fame and to reputation up to been considered with the paintbrush. among them, Miquel Bestard, " the Majorcan " (1592-1633)[36], "Lucas de Landa "[37], the famous painter and Aragonese jurist Micer Paul (1593-1659)[38], "Espadarino"[39], Francisco Collantes (1599-1656) of Madrid[40], Rafael Pertus (1567-1648), Luís de Vargas (1505-1567), Sevillian painter who resided many years in Italy, and the Zaragozano Pedro Urzanqui. After all, the circuit for the acquisition of pictures by Lastanosa appears us complicated, though we know the existence of other similar circuits in other areas, specially the international ones, based on commercial nets and many of them established? under the arrival of the explorers, missionaries, merchants, agents, and Europeans fortune-hunters to all corners of the globe. One of these agents brought from France a statue in bronze of Neptune that for Lastanosa that had belonged to his "very special friend," the Count of Guimerá, Gaspar Galcerán de Castro y Pinos (1584-1638)——great scholar, collector of ancient coins, and pioneer among antiquarians of Aragon.[41] Vincencio Juan de Lastanosa inherited from the Count of Guimerá several manuscripts as well:

> Estatua de bronce de Neptuno sobre pedestal de evano, que se compro en la ciudad de Nimes en Francia de un cavallero que la tenia en su casa desde el saco que dieron los franceses a Florencia año [...], y deve España el goçar esta preciosa antiguedad a la grandeza del ilustrissimo señor don Gaspar Galceran de Gurrea y Aragon Castro y Pinos, Conde /[f. 44r.] de Guimera y Vizconde de Evol y Alquer Foradat, ornamento ilustre de los antiquarios, el qual embio un gentilhombre de su casa a Francia solo para las ferias desta estatua.[42]

The relation of Lastanosa with well-known prominent figures of the Aragonese power does not stop here. Lastanosa had a portrait of Diego de Arnedo, bishop of Huesca and holder of another collection of ancient coins[43]. Collecting coins was an interest that also had another personage whose Lastanosa had a manuscript, Philip Puivecino and Castro[44]. Lastanosa also received donations from D. Fernando Manuel of Aragon de Gurrea and de Borja (ex-De Borja and Aragon), VIII Duke of Villahermosa, VII Count of Spanish Parliament and III of Ficallo, Mr. of Gurrea's barony (d. 1664?). Beyond a " immediate circle ", nearby in distance, agile for proximity, but, simultaneously, not still the being sufficient to be able to sustain in him the cause of everything assembled by Lastanosa, we see the existence of a set of slightly more external relations. These relations were as interesting and important as the previous ones and, in many cases, similar, both for its content and for the form. Being them Aragonese or not, sometimes probably before nearby and sometimes, distant, Lastanosa was currently kept to of the novelties in the matters he was interested, by the mere fact of acquiring a book, or of that receive one as a present. We have examples for both cases. In the first one, we see that he had, among his printed books:

> "De la Hidrografia: varias cartas de marear antiguas y modernas. De la enmendacion[45] de ellas, al Excelentissimo Señor D. Pedro Porter y Casanete, Almirante de las Califurnias y de presente Visorey del Reyno de Chile."[46]

In the second one, the Count of San Clemente:

> Olivari Uredi I.C., Sigilla comitum Flandriae. En folio, Burgis Flandorum 1639[47]. Con este tomo enriqueció mi libreria el Illustre Señor Conde de San Clemente[48], cavallero en quien se logran nobleza, virtud, afabilidad y agasajo con tantas ventajas que puede ser dechado de todas las buenas prendas, realzandole mucho la eleccion de los buenos libros con que cada dia aumenta su museo y la de pinturas con que illustra su palacio".[49]

A relevant fact for knowing the formation and aggrandizement of the Lastanosa library and collections was his trip to Madrid in 1676, where he met many booksellers and other collectors, acquiring pieces and texts everywhere. One more time, a new group of relations appear around Lastanosa, even when his appointments did not were only for enriching his patrimony. We know the information thanks to a letter sent from Huesca, when Lastanosa return of Madrid. The letter was sent to the Zaragoza nobleman Miguel Martin de Villanueva y Palafox, Count of San Clemente, and that we can read transcribed thanks to Carlos Garcés Manau[50]. In any case, he entertained himself in "speaking with virtuous men", among which were friar Hermenegildo de San Pablo and Jose Pellicer de Ossau y Tovar (1602-1679). I mention them here joined in respecting the relation that both had in many moments of his life. Specially well-known are the Jose Pellicer's economic difficulties after marrying for third time with Isabel Maria Ogacio de la Torre, moment in that Fr. Hermenegildo de San Pablo was giving to them, as charitable gift a fish and two birds[51]. The list of prominent figures that Lastanosa saw in Madrid is long and there includes many scholars and holders of big libraries, as Gaspar Ibáñez of Segovia (1628-1708), Marquess of Agrópoli, whose interest to the books was making him to visit frequently to Jose Pellicer. In fact, Gaspar Ibáñez would entrust Pellicer a familiar anthology, genealogy de mode, Like Lastanosa forged Lastanosa, which appeared entitled *Memorial de la Casa y Servicios de don Gaspar Ibáñez de Segovia y Peralta, Cavallero del Orden de Alcántara, Señor de la Villa de Corpa, al Rey Nuestro Señor*[52]. Lastanosa also met with "eclesiastico muy erudito, mui curioso correspondiente de don Nicolas Antonio, Agente del Rey de España en Roma. Presentome una mapa de Madrid, unos mariscos mui raros para una gruta, y los dos tomos de la Biblioteca Hispanica[53], que se los havia remitido el autor para que me los imbiara a Huesca". This saged cleric man was no other than licentiate Blas Canales de Carranza, major Chaplain of the Congregation of Our Father San Pedro of Presbiters Priests of Madrid, in San Pedro's church, and who signed on May 20, 1663 as insider of Pedro Calderón de la Barca (1600-1681). Not less interesting is his interview with Juan de Mendoza, Weapon Major of King of Philip IV, who gave to him "all his nobiliaries"[54]. Once visited these persons, Vincencio Juan de Lastanosa passed to do a tour visiting the booksellers of Madrid. At in this moment, we see how well Lastanosa organized and programmed his travel to Madrid. May be that even he had coordinated before the interviews, because all of them have a point jointly: of one or another way, all the booksellers have relation with Huesca or with Aragon. One of the first ones who visited was José de Alfay. Bookseller Alfay did not gave a good account of himself to Lastanosa, since he qualifies him of " dry, greedily, inaccessibly ". Lastanosa wanted to buy him them for the Count of San Clemente " "seis tomos del Tostado en romance en un doblon de a 8, con animo de servir a V.S. con ellos"[55]. Alfay was an active merchant of books, publishing and printer, that came from Saragossa to live in Madrid. He never lost his contacts at Saragossa[56], specially with his father and uncle, also booksellers. Alfay even managed to praise the Lastanosa's alchemist qualities[57]. His *Poesías varias* suppose the culmination of the whole anthology, and was dedicated the illustrious poet of Tortosa Francisco de la Torre (1625-1680). This writer, who died being the most famous of his times in Catalonia, considered the Lastanosa collections as element of motive in his works.

Lastanosa followed with the booksellers and the shopping:

> Estube con Juan Anison, librero frances. Tiene buenos libros, es galante, pero carissimo. Ad [...]bien es mas tratable, noticioso, he quedado en su gracia, y espero conseguir con su comunicacion algunos buenos libros. Lavastida tiene muchos libros. Hele comprado dos tomos de Medicina de un inglés para el Protomedico, que se los inbiaré en haver ocasion. De 13 ó 14 tomos en 4, que son las obras de Escoto de seis qu ... tiene. Le dava 16 reales de a 8. Hoi no los ha querido dar, he dejado de comprarle otros libros curiosos por la caristia. Juan de San Vicente, aragones, es mui caudaloso, caro y rustico ..."

We see, that among them they were the Frenchman "Juan Anison"[58], "Labastida"[59], and very nearby to the previous one, the Aragonese Juan de San Vicente. These last two booksellers, beyond their importance as professionals, had a narrow relation with Lastanosa. The first one of them, Mateo de La Bastida was coming stamping Baltasar Gracián's works from several years before the Lastanosa travel to Madrid. And Lastanosa was the patron of Baltasar Gracián, from 1663[60], managing to publish two years before El Discreto[61].

We have done an incomplete tour on some facts in the life of Vincencio Juan of Lastanosa, more punctual and different some less. We have left behind a deeper analysis of his relations with other prominent figures, that might help to define better Lastanosa as collector, lover of the numismatic, of the books, of the plants and, why not?, lover of the relations in that he was interested, as eternal aspirant to a social very important category in his time, that is nobleman, who never managed. This, nevertheless, did not prevent him exercising until the perfection a role, a proper role, probably designed by him, probably imposed by the circumstances. A group of circumstances, so much close as those of the years in which he lived or that they forced by his past. 17th century Spain is simply a spectacular century, capable of monopolizing the attention of the investigator for all his life. Did Lastanosa obtain his objective? What does it was? We do not know it. Or probably yes. It can be that he always was imitating someone, or can be also that everything what was doing was approaching inside the normality "of theirs ", for what I incline. We can see him since whatever we want.

Notes

1. Maurice Caillet, & Robert Mesuret, « François Filhol, toulousain, son oeuvre et son cabinet (1583?-1648) », *Mémoires de la Société Archéologique du Midi de la France*, XXIX (1963), 99-137, Adolphe Coster, « Antiquaires d'autrefois. A propos de quelques lettres inédites de François Filhol, hebdomadier de Saint-Etienne de Toulouse, au chroniqueur d'Aragon Don Francisco Ximénez de Urrea », *Revue des Pyrénées*, XXIII (1911), 436-471.
2. The description of François Filhol's collection was done in Spanish by Juan Francisco Andrés de Uztarroz and for Lastanosa in *Diseño de la insigne i copiosa Bibliotheca de Francisco Filhol*, (Huesca, Juan Francisco de Larumbe, 1644). In vice versa, Juan Francisco Andrés de Uztarroz himself, also wrote the description of the Lastanosa's collection for Filhol in *Descripcion de las Antiguedades y Jardines de Don Vincencio Iuan de Lastanosa*, (Zaragoza, Diego Dormer, 1647). The first one is in verse and it is a copy, in a free translation of the manuscript that Filhol wrote, of not more than four pages, for Gaston de Orleans (1608-1660), the third son of the French King Enrique IV (France, Bibliothèque Nationale, in forward BNF, manuscript français 390: " François Filhol's Écrits divers, hebdomadier in l'eglise metropolitaine of Toulouse ", Part 2, 5-6v: "Description of his cabinet"). May be Ustarroz could have done this copy in the city of Huesca, which it implies that a copy of the manuscript was sent by Filhol to Uztarroz, or to Lastanosa. Even it is possible that Ustarroz could travel to Toulouse. The direct comparison of both texts allows us to assure that Uztarroz followed the manuscript, though we know if the differences are due to a mistake or not, since it

happens when there is a difference of thousand books less in the translation to the Castilian language:
« Ledit cabinet est une grande salle environee depuis le carriau jusques au plancher de plus de trois mille volumes de livres […]. » BNF, manuscrito français 390, 6.
« Entrase en otro salón, cuyas paredes del suelo al techo están rodeadas de estantes, y en ellas hay más de dos mil libros». Biblioteca Nacional de Madrid, in forward BN, 11318/13: "Diseño de la insigne, y copiosa biblioteca de Francisco Filhol", 4. José Simón Díaz, *Bibliografía de la Literatura Hispánica*, Madrid, CSIC, 16 vols., Vol. V (1958), 405, 2650.
Nevertheless, we insist that what more prevails, so much in the manuscript text since in Ustarroz form it is the loyalty:
«Montant le degré, nos amis y rencontrent les armes antiques de diverses nations […] » BNF, manuscrit français 390, 5v.
"Subida la escala se ofrecen a la vista armas antiguas de viarias naciones." BN, 11318/13, 3.
On Uztarroz : Ricardo del Arco, *La erudición española en el siglo XVII y el Cronista de Aragón Andrés de Uztarroz*, Madrid, C.S.I.C., 1950.
3. "Don Francisco Filhol, lustre, ornamento y elogio de la ciudad de Tolosa" (Uztarroz, Andrés de, *Diseño de la insigne*, 11). "El docto Francisco Filhol, celebre por sus letras y por el gran numero de libros y cosas raras de la naturaleza, medallas y estatuas que juntó en su casa en Tolosa, en el claustro de S. Cernil" (*Narración de lo que le pasó a don Vincencio Lastanosa a 15 de octubre del año 1662 con un Religioso docto y grave*, Hispanic Society, ms. B-2424, 66v.
4. "Otro libro en folio de Ecechiel en pergamino, con caracteres y figuras doradas y coloridas con notable prolixidad. Fue dadiva del Sr. Emperador Carlos Quinto a un erudito y noble frances, y a mi libreria lo franqueó el docto Francisco Filhol…" (*Narración de lo que le pasó*, 66v).
5. This church is called in the manuscript "S. Cernil " , in a bad transcription of the manuscripts. Nowadays this church is the <Magnificent basilica of Toulouse>. Built in the century XI, in the "Way of Santiago", the church is a singing to the Romanesque Art. The cloister and the lateral abbey both disappeared when it was dismantled between 1804 and 1808, and due to its progressive abandon after the French Revolution and the desire to give a better access to all the doors of the church. Nevertheless, we still can see some of its capitals in the Museum of the s. Agustin Order at Toulouse. This Church becomes Cathedral in 1878.
6. BNF manuscript abstract concerning Filhol and his gardens:
« Premièrement il a sa salle basse tapissée de tableaux et environnée de tous costez de plusieurs instrumens de musique, tant anciens que modernes, avec une grande quantité, de livres ou parties completes pour cet art. Montant le degre, nos amis y rencontrent les armes antiques de diverses nations […] Le dit cabinet est une grande salle environee depuis le carriau jusques au plancher de plus de trois mille volumes de livres […] L'estendùe de le veüe y propose d'abord trois rangs de petits portraits […] le tout parseme de plus de quatre fines platines de plom, moulées d'apprez argent ciselé en histoyres, fables, ou grotesques d'inventions. Les vuides

pour la distance des livres sont remplis de petits païsages de chasse, et de vases de fleurs. Du coste de l'Occident nous avons le Cabinet des médailles a cens tiroirs, couvert d'un velin rouge dore a petit fer en nombre de quinze a seze cens pièces [...] La cassette des camayeux et de la pierrerie est à main droite en nombre de pres de plus de six cens pieces dont la plus grande partie contient les sigilles des Romains, ou anneaux des Payens, devises, emblèmes, énigmes historiques et fabuleuses a la graveure antique [...] Du coste du midy [...] il y a une table longue de plus de deux canes et demy et large de six pans entièrement couverte et surchargee de plus de mille coquilles diverses en espece et grosseur ou couleur extraordinairement curieuses. De là qui est le seul espace vuide pour se placer, on void les tables ajustees par certains degrez de proportion qui remplissent la chambre du côté de l'aquilon, couvertes des prodiges de l'air, de la mer et de la terre, comme oyseaux, poissons, animaux, anatomies dans l'esloignement, et au plus proche dela veüe un petit sepulchre d'albastre très rare, les urnes en cuivre, de terre et d'argile, les larmoyres diverses, les lampes corinthiques de fer et de terre, les vases de la sacrificateure des Anciens, les statues de bronze, les testes de marbre, les figures d'yvoire, les cachets, escussons, bagues, poids, clef, serreures antiques et les dieux des Egyptiens. Jusques aux Priapes antiques. Passant à une autre salle par deux petits cabinets, le premier est remply des outils servant à la mécanique en grande quantité, avec les merveilles ou chef-d'œuve de plusieurs artisans. Le second contient l'assemblage des raretez de la sculpture, ou nous avons les troncs du Bandinel, les corps de Michel-Ange; les postures de Biard, les testes de Bachelier, les pieds et les mains des plus experts en cet estude [...] Il y a encore une table alignee a proportion couverte de quinze douzaines de grandes coquilles de nacre de perle remplies de rocaille, de petrifications ou empierremens admirables, de coraux divers, de mineraux et fossilles, les cailloux des pierreries bruttes, les matrices des diamans de Zeilan, le jaspe, le porphire, trois piramides de marbre avec une grande quantite de boules de diverses couleurs. Nous avons ensuite la cassette des coleurs pour nostre laveure et miniateure, le coffret des platines d'esmail en nombe de septante et quatre cens pièces ou sont les douze Cesars touchez d'une excellente main et bien différens de ceux qu'on nous apporte aujourd'huy de Limoges. Le coffret des papillons divers est a l'opposite en nombe de plus de deux cens pièces, objet ravissant pour louer et admirer par coes creaieures les merveilles du Createur [...] Outre ces merveilles en gros, nous avons encore de surcroist les chefs d'œuvre d'un horlogeur autant inimitable qu'il est incomparable en un pinacle de cuive doré, aux febestrages d'argent, fondé sur un hexagone [...] »

7. Andrés de Uztarroz, *Descripcion de las antiguedades*, 15.

8. *"Y en el espacio que ay entre el desaguadero y la calle principal y el estanque se admira un jardin, cuyas divisiones son de boxes, y en el logran la vista y el olfato desde el tulipan hasta el junquillo"*. Andrés de Uztarroz, "Descripción del palacio y los jardines de Vincencio Juan de Lastanosa", Hispanic Society, ms. B-2424, 50v.

9. Peter M. Garber, "Tulipmania", *The Journal of Political Economy*, 97 (1989), 535-560. Anne Goldgar, *Tulipmania*: *Money, Honor, and Knowledge in the Dutch Golden Age*, Chicago, The University of Chicago Press, 2007.
10. Pierre Borel, *Antiquitez, raretez, plantes, mineraux, et autres choses considerables de la Ville, et Comté de Castres d'Albigeois*, Castres, Arnaud Colomiez, 1649, 124-131 : « *Roole des principaux cabinets curieux, et autres choses remarquables qui se voyent ez principales Villes de l'Europe* ».
11. « Mr. Me N. de Puimisson, Conseiller, curieux de medailles, et de tableaux. Le cabinet de François Filhol, Hebdomadier de l'Eglise Sainct Estienne. Mr. Clemens, Chanoine de S. Estienne, curieux des fleurs. Mr. Catel Official. Mr. Paucy, Conseiller. Mr. de S. Ipoly, pour les esmaux anciens. Mr. Roc. Le cabinet de feu Domairon, Cordelier. La Bibliotheque enchaisnée des Cordeliers. Mr. de la Bourgade, Chanoine. Mr. Dirat, Sacristain de S. Estienne. Mr. Charles Pradié. Mr. Boutonnier. Mr. de Fresals, Conseiller Clerc. Mr. du May. Mr. de la Combe, Audiencier à la grand Chambre, et Mr. Nicolas Choisy, marchand. » Pierre Borel, *Antiquitez,* 124.
12. Pierre Borel, *Bibliotheca chemica seu catalogus librorum, philosophorum, hermeticorum ...usque ad annum 1643*, París, Carolus du Mensil, 1654.
13. "The duke of Orleans sent, among other things, several salted and dried animals and he requests in 1631 drawings of several statues of his collection". José Morán & Fernando Checa, *El coleccionismo en España. De la cámara de las maravillas a la galería de pinturas*, Madrid, Cátedra, 1985, 197. Reference from BN, manuscript 18.727-45 (*Las tres cosas mas singulares que tiene la Casa de Lastanosa en este año de 1639*). This manuscript is actually seen as a false source by Fermín Gil Encabo. See Fermín Gil Encabo, "La ficción "telamoniana" de Pellicer en torno a Lastanosa", *Actas del V Congreso de la Asociación Internacional Siglo de Oro Münster, 1999*, Madrid, 2001, 623-634. Jacob Soll, "Empirical History and the Transformation of Political Criticism in France from Bodin to Bayle", *Journal of the History of Ideas*, 64- 2 (2003), 297-316.
14. Fortunately, the words of some nineteenth-century Spanish historians pushing back that the collections had some interest are very far:
"Pero como de todo el hombre abusa, entre tantas y curiosas colecciones como en el siglo XVII se efectuaron, hay muchas que están muy distantes de presentar un verdadero provecho e interés. No hablamos solamente del que puedan excitar los semblantes de ínclitos varones, sino que también del de varios objetos secundarios que dan á todo esto cierta importancia y utilidad, cual es el conocimiento de los trajes, armaduras, muebles y otras usanzas de épocas remotas. Empero el orgullo del hombre ha querido hacer vana pompa y ostentación de antiguos é ilustres ascendientes [...] Pero éstas son honrosas excepciones que nada tienen que ver con los abusos y largas falsificaciones de retratos, sin otro objeto que halagar el orgullo y la vanidad". Valentín Carderera y Solano, "Ensayo histórico sobre los retratos de hombres célebres desde el siglo XIII hasta el XVIII, el origen de sus colecciones en Europa, particularmente en Italia y en España, y examen crítico sobre su autenticidad y la de las numerosas colecciones grabadas desde fines del siglo XV hasta nuestros días, leído por su autor D. Valentín Carderera a la Real Academia de

la Historia el 19 de abril de 1841", *Boletín de la Real Academia de la Historia*, 34 (1899), 201-257, especially 243-245.
15. Miguel López Pérez, "Anatomía del virtuoso: coleccionismo y melancolía en la figura de Juan Vicencio de Lastanosa", *Argensola*, 115 (2006), 143-164.
16. Jean Baptiste Dru, *Catalogue des plantes, tant des Tulipes que des autres fleurs qui se trouvent au jardin du Sieur Jean-Baptiste Dru, Herboriste du Roy, demeurant proche la Déserte à Lyon. Troisième édition, augmentée de quantité de plantes, & des noms d'une partie des couleurs qui manquoient en la precedente impresión*, Lyon, Guillaume Barbier, 1653.
17. Biblioteca Nacional de Madrid (BN) 18727-55.
18. M. F. Wamer, "The Morins", *National Horticultural Magazine*, 33 (1954), 168–176.
19. « Le lendemain matin, un ami me counduisit au jardin de M[onsieur] Morin, qui, de simple jardinier, est devenu l'un des plus savants & plus habiles collecteurs de fleurs, de coquilles & d'insectes. Le jardin es de forme parfaitement ovale & entouré de cyprès palissadés & taillés, aussi unis que si c'éttoit un mur. Ses tulipes, ses anémones, ses renoncules, ses crocus, & c., sont des plus rares, & attirent chez lui leurs admirateurs durant tout le saison. Il habitoit un espèce de l'hermitage à un bout de son jardin, au milieu de ses collections de coquilles & des coraux, dans l'un desquels on a taillé un grand cricifix qu'on estime fort. Il a aussi des livres d'estampes d'Albert Durer, de Vankeyden, de Callot, &c. Sa collection d'insectes, sourtout celle de papillons, est des plus curieuses : une préparation qu'il fait subir à ceux-ci les rend incorruptibles, & ils sont arrangés sur des cartons de façon à représenter une belle tapisserie. » Martin Lister, *Voyage de Lister à Paris en 1698*, París, Société des Bibliophiles, 187, 258-259.
« Il est vray qu'il se voit de coquilles rares & mervelleusement diversifiées, dont sont ornez aujourd'huy tant de Cabinet curieux. Et sour tout celuy de Monsieur de Monmot Maistre de Requestes, où cet excellent homme les delices des Muses, & l'amour de toutes les belles Ames, en a recueilli de tant d'especes diferentes, auissi-bien que nostre virtueux Ami Monsieur Morin, si versé dans les connoissances des plantes, & qui ne prescrit point de bornes à ses curiositez. L'y en ay veu qui portent les perles, & quelques autres dont les Anciens tiroient cette pourpre precieuse… » Michel de Marolles, *Suite des mémoires de Michel de Marolles, Abbé de Ville-Loin, contenant douze traitez sur divers sujets curieux...*, París, Antoine de Sommaville, 1657, 95.
20. Vincenzo Mariscotti, *Alla Potentissima Sacra Caesarea Maestà de Leopoldo Ignatio d'Austria Imperatore, Sempre Pio, sempre Felice, sempre Augusto, Invito à gl'incontri della Reale Augusta sua Sposa Margherita d'Austria Infanta delle Spagne, Ode del Conte Vincenzo Mariscotti di Bologna*, Bologna, Gio. Battista Ferroni, 1666.
21. *Narración de lo que le pasó*, 78-78v.
22. Johns Adrian, *The Nature of the Book: Print and Knowledge in the Making*, Chicago, Chicago University Press, 1998.

23. Daniela Bleichmar, "Los usos del libro: La circulación de conocimiento en la Edad Moderna", in: Rey Bueno, Mar (ed) *La Bibliotheca Magica.* Madrid, Servicio de Publicaciones de la Universidad Complutense de Madrid, 2007, 33-38.
24. Paolo Maria Terzago, *Musaeum Septalianum Manfredi Septalae Patrit Mediolanensis*, Dertonae, 1664. Scarabelli, P.. F., *Museo ò galeria adunata dal sapere, e dallo studio del sig. canonico Manfredo Settala nobile milanese. descritta in latino dal sig. dott. fis. coll. Paolo Maria Terzago et hora in italiano dal sig. Pietro Francesco Scarabelli. Dott. fis. di Voghera. E dal medemo accresciuta*, Tortona, li figliuoli del qd. Eliseo Viola, 1666.
25. Bernard de Montfaucon, *Diarium italicum, sive Monumentorum veterum, bibliothecarum, musaeorum, &c. notitiae singulares in itinerario italico collectae, additis schematibus ac figuris / a R. P. D. Bernardo de Montfaucon,...*, Paris, Joannem Anisson, 1702.
26. Johann Jacob Dillen, *Hortus Elthamensis seu Plantarum rariorum quas in horto suo Elthami in cantio coluit... Jacobus Sherard,... delineationes et descriptiones quarum historia vel plane non, vel imperfecte a rei herbariae scriptoribus tradita fuit,* Londini, J. J. Dilleno, 1732 ; Tobias Aldini Cesenate, *Exactissima descriptio rariorum quarundam plantarum quae continentur Romae in horto Farnesiano / Tobia Aldino Cesenate auctore*, Roma, J. Mascardi, 1625.
27. Ludovico Settala, *Ludovici Septalii patricii et medici Mediolanensis ... De ratione instituendae, & gubernandae familiae : ibri quinque / senator F. edidit, & Iulio Aresio ... dicauit*, Mediolani, Io. Baptistam Bidellium, 1626.
28. Vidania Introduction to Vicencio Juan de Lastanosa, *Tratado de la moneda jaquesa y de otras de oro y plata del Reino de Aragón*, Huesca, 1681.
29. Carla Tavernari, "Manfredo Settala, collezionista e scienziato milanese del '600", *Annali dell'Istituto e Museo di Storia della Scienza*, 1 (1976), 43-61.
30. Philippus Picinelli, *Mundus symbolicus in emblematum universitate formatus, explicatus, et tam sacris, qum profanis eruditionibus ac sententiis illustratus... /Reverendisimo Domino D. Philippo Picinello...*, Coloniae Agrippinae, Sumptibus Hermanni Demen, 1687.
31. Nowadays, Settala collection is partially disappeared. The rest is stored in Milan (Sforza Family Castle and Science and Technique Museum.
32. Codici Ambrosiani Z 387–388–389 sup., Museo Settala (inventario degli oggetti, con disegni), voll. 3 Cart., mm. 310x220, cc. III+56+I, cc., sec. XVII (1640/'60).
33. Vincencio Juan de Lastanosa had some busts and portraits. Among them, the bust of his alchemist Nadal Baronio and the portrait of Juana Navarra y Rocafull, Vice-Count of Torresecas, daughter of Giner Rebassa de Perellós and María de Rocafull, Vice-Count of Torresecas. The Vice-Count of Torresecas was stablished the march 18, 1678 to Francisco Cortés y Sangüesa (Andrés de Uztarroz, *Descripción del palacio*, 32v).
34. Padilla Manrique y Acuña, Luisa Maria de, Condesa de Aranda (ca. 1590-1646), *Idea de nobles y sus desempeños en aforismos: parte quarta de Nobleza virtuosa*, Zaragoza, en el Hospital Real y General de Nuestra Señora de Gracia, 1644; Adolfo Carrasco, "Los Mendoza y lo sagrado. Piedad y símbolo religioso en

la cultura nobiliaria", *Cuadernos de Historia Moderna*, 25 (2000), 233-269. Adolfo Carrasco, "Fisonomía de la virtud: Gestos, movimientos y palabras en la cultura cortesano-aristocrática del siglo XVII", *Reales sitios: Revista del Patrimonio Nacional*, 147 (2001), 26-37.
35. The number of counts, dukes, marquis and viscounts was continuously increased by the Spanish Kings. 55 titles with Charles I, 94 with Philip II, 144 with Philip III, 236 with Philip IV, and 536 with Charles II.
36. Andrés de Uztarroz, *Descripción del palacio*, 25, with name of "Vestart". Miquel Carbonell i Buades, "El pintor Miquel Bestard (1592-1633), el Mallorquí; notícies biogràfiques i aportacions al catàleg", *Locus Amoenus*, 2 (1996), 155-174.
37. Ricardo del Arco mentions as painter's name "Lucas de Holanda" (Ricardo del Arco y Garay, R., *La erudición aragonesa del siglo XVII en torno a Lastanosa*, Madrid, 1934). He was Lucas van Heere (1534-1584), one of the favourite's painters of Queen Elizabeth I of England.
38. Religious themes painter, his Descent from the Cross was possesed by Marquis of Ayerbe (Pedro-Jerónimo Jordán de Urríes y Arbea y Agustín, d. 1669) and Marquis of de San Clemente. In august 31, 1620, an *Ecce Homo* of Micer Pablo costs 1.600 coins:
"Juan Sanz de Armora, doctor en derecho y Nicolás de Cascarosa, notario, promulgan una sentencia arbitral por las diferencias habidas entre Silvestre Estanmolin, pintor, vecino de Zaragoza, e Isidoro Domingo Cortés, doctor en medicina, domiciliado en dicha ciudad, y heredero universal de la difunta Matías Cortés, tía suya y mujer de dicho Silvestre Estanmolin, sobre algunos bienes muebles pertenecientes a la causante, y especialmente acerca de un cuadro de un Ecce Homo, pintado al óleo por micer Pablo, y tasado por los arbitros en 1.600 sueldos."
Ana Isabel Bruñén Ibáñez., Luis Julve Larraz, & Esperanza Velasco de la Peña (ed.), *Las Artes en Aragón en el siglo XVII según el Archivo de Protocolos Notariales de Zaragoza. De 1613 a 1696*, Zaragoza, Institución «Fernando el Católico» (CSIC), v.III, (2005), 165.
39. The italian painter Giacomo Galli, "Spadarino" (1585-1651).
40. Vicente Carducho' disciple (1576-1638).
41. Carmen Morte, "Emblemas en un manuscrito aragonés del siglo XVII: "Honestas recreaciones de las monedas y medallas" del Conde de Guimerá", *Emblemata*, 9 (2003), 315-382.
42. Andrés de Uztarroz, *Descripción del palacio*, 44-44r.
43. Andrés de Uztarroz, *Descripción del palacio*, 30. Arrendó was Bishop of Huesca between 1572 and 1574, following his episcopate in Palma de Mallorca (1561-1572) as well.
44. "Un libro manuscrito del Doctor Felipe Puivecino, Dean y Canonigo de la Seo de Huesca, insigne por sus letras y benemerito de la venerable antigüedad", in *Narración de lo que le pasó*, 52v. Puivecino y Castro (154?-162?) was vicar-general of the Huesca diocese in the year 1570. He had a big cabinet full of medals and ancient coins, just like said Francisco Diego de Aynsa e Iriarte (1586-1628), in *Fundación, excelencias y grandezas y cosas memorables de la*

antiquísima ciudad de Huesca (Imp. Pedro Cabarte, Huesca, 1619, 16, col. 1). See also Gerónimo Pujades (1568-1635), *Crónica universal del principado de Cataluña*, Barcelona, Hieronym Margarit, 1609, lib. 3, chap. 91, 140.

45. Ricardo del Arco y Garay incorrectly wrote "encomendación" (Ricardo del Arco y Garay, *La erudición aragonesa*, 1934). Really, is right the word "enmendación" of navigation chart. The navigation charts emendation was a repeating problem that came face to face with pilots, cosmographers, mapmakers and others theorists, especially during 16th century, after the rising of sea circulation to West Indies. And also, this problem provoked hard controversies among them. Alison Sandman, "An Apologia for the Pilots' Charts: Politics, Projections and Pilots' Reports in Early Modern Spain", *Imago Mundi*, 56 (2004), 7 - 22 and *Pilots versus cosmographers. Education and the practice of navigation in 16th-century Spain*, University of de Wisconsin, unpublished PhD Dissertation, 2001. Francisco Fernández y González, Francisco (1833-1917), "Graduaciones náuticas de las Cartas de Indias", *Boletín de la Real Academia de la Historia,* tomo 4 (1884), 7-159.

46. In reference to Pedro Porter Casanate, (1611-1662), *Reparo a errores de la navegación española al excellentissimo señor Don Fadrique de Toledo*, Zaragoza, María de la Torre Ossorio, 1634. *Narración de lo que le pasó*, 55.

47. In reference to the first of five tomes in folio of the immense work on the History of Flanders that wrote the jurist Olivier de Wree (1596-1652), *Sigilla comitum Flandriae*, Brugis Flandrorum, 1639-1650. At the same time de Wree was patron of Lambertus Vossius (Lambert de Voos) (¿-1661).

48. He had, about 1680, an excellent collection and library at Zaragoza, including very rare texts, such as the codex *Libro del Conosçimiento de todos los rregnos et tierras et señorios que son por el mundo, et de las señales et armas que han*, Münich, Bayerische Staatsbibliothek, Cod. hisp.150. County of San Clemente started by confernment of King Philip IV to Juan Martínez de Villamesa in march 24, 1640.

49. *Narración de lo que le pasó*, 71v.

50. Carlos Garcés Manau,, "Un Lastanosa poco conocido (1665-1679). Las relaciones con Juan José de Austria", *Argensola*, 115 (2005), 41-93, letter translated in 84-85.

51. Biblioteca Nacional de Madrid, ms 8.388, 387, year 1672. On these questions: Juan Manuel Oliver, "Los matrimonios de José Pellicer (Noticias de su vida familiar y descendencia)", *Criticón*, 63 (1995), 47-88. on relations between Pellicer and Vincencio Juan de Lastanosa: Fermín Gil Encabo, F., "La ficción "telamoniana", 623-634. Fray Hermenegildo de San Pablo (O. Jer.), *Instruccion previa a los lectores de la instruccion historica apologetica del padre Fr. Gregorio Argaiz, benedictino, y examen del crisol purificativo del padreFray Manuel Leal, augustiniano, por Fr. Hermenegildo de San Pablo del real convento de San Geronimo...*, Zaragoza, Herederos de Diego Dormer, 1676; Fray Hermenegildo de San Pablo (O. Jer.), *Origen y continuacion de el instituto y religion geronimiana fundados en los conventos de Belen en Palestina por San Geronimo, explayados a los monges geronimos de España, aberiguados por Fr.*

Hermenegildo de San Pablo del real conuento de San Geronimo de Madrid, Madrid, Imprenta Real, 1669; Fray Hermenegildo de San Pablo (O. Jer.), *Defensa por la Religion Geronyma de España, y su antiguedad : en que se responde a vn tratado, que el autor de la Poblacion eclesiastica imprimiô en su quarta parte, el año de mil seiscientos sesenta y nueve, contra el origen geronymiano / responde a la impugnacion, y a su forma, el Padre Fr. Hermenegildo de S. Pablo, hijo del Real Convento de S. Geronymo de Madrid* Zaragoza, imprenta de Diego Dormer, 1672. In this last book, Hermenegildo de San Pablo tried to decrease the power of the Carmelitan Order in Spain, receiving a lot of excessive criticisms, because:
"Con tanto deslustre de las demás Religiones pretende introducir la suya en el mundo, no teniendo de Santo más que el nombre [...], como si el mayor lustre de la suya, no estuviera en dar a las demás Religiones, lo que se les debe de justicia". Fray Jerónimo de La Concepción (O.Carm.), *Emporio del Orbe. Cádiz Ilustrada*, Ámsterdam, 1690, III, c. VI, 18 and IV, c. XIV, 11.
Nevertheless, he was extolled by Marcelino Menéndez Pelayo:
"...no sin que, para gloria de la crítica histórica entre nosotros, encontrasen los osados falsarios la formidable oposición de varones tan preclaros como D. Juan Bautista Pérez, Pedro de Valencia, Fr. Hermenegildo de San Pablo, el marqués de Mondéjar, D. Juan Lucas Cortés y D. Nicolás Antonio."
Marcelino Menéndez Pelayo, "Indicaciones sobre la actividad intelectual de España en los tres últimos siglos. Carta al Sr. D. Gumersindo Laverde Ruiz, Catedrático de Literatura en la Universidad de Valladolid, &c.", *Revista europea*, 117 (1876), 330-339, reference in 337.
52. Madrid, 1649.
53. Nicoás Antonio, *Bibliotheca Hispana vetus sive Hispanorum qui usquam unquámve scripto aliquid consignaverunt notitia... : tomus primus [-secundus] auctore D. Nicolao Antonio... ; opus postumum*, Roma, Typographia Antonii de Rubeis, 1696
54. As Weapons Major, Juan de Mendoza was verifying the lineage before giving nobility grace by the King. Juan de Mendoza, *Minutas de linajes de España de D. Juan de Mendoza, Rey de armas de Felipe IV y Carlos II*, Madrid, 1650-1699. Juan de Mendoza, *Apuntamietos genealógicos*, Madrid, 1650.
55. In reference to Alfonso (or Alonso) Fernández de Madrigal, known as "el Tostado" (1400-1455) and his work *Eusebio de las crónicas o tiempos* (1445-1450). *Eusebio* was printed in Salamanca (H. Gysser, 1506-1507), in six volumes. Lastanosa was interested in this edition, but not in the Latin printed edition of thirteen volumes at Venetia (1507-1531).
56. His ather was bookseller at Huesca: Jerónimo de Blancas, *Coronaciones de los Serenissimos Reyes de Aragon escritas por Geronimo de Blancas, con dos tratados del Modo de tener Cortes, publicalo Iuan Francisco Andres de Uztarroz, con algunas notas*, Zaragoza, Diego Dormer, a costa de Pedro y Tomas Alfay, 1641. See also Caros Garcés Manau, "Un Lastanosa", 41-93, footnote 20.
57. Carlos Garcés Manau, "Un Lastanosa poco conocido, 49-50. Alfay was the author of the text *Sarao de Aranjuez de varios versos y novelas* (Madrid, 1666, BN, R/1169), dedicated to Lastanosa:

"El Barón y Mecenas de las Letras, el imán de todas Ciencias, la gala de la Cultura, la feria de lo curioso, para cuyo gusto compitieron la Naturaleza a prodigios y el Arte a marabillas. Véase en esse culto Camarín de las Musas, Museo de quanto precioso estimó el Orbe y oy venera reservado de la guadaña voraz del tiempo, a la erudición de los siglos en pinturas de eminentes maneras, en monedas y estatuas de venerable antigüedad, dexando en el mismo Camarín a los venideros fuentes de salud e ideas para hazer tesoros de Medecina, como dezía Damasceno, del oro y perlas, que hasta agora no sabía más su amor que sepultarlos, pero su industria y estudio de V. m. los muestra líquidos, potables, para que su uso aparte por muchos lustros a los mortales del sepulcro. A cifrado al fin V. m. en su Casa el Panteón Romano, pues todo lo goza junto y lo participa a los demás. Las delicias de los Iardines de Floras, Pomanas, Elisios. Las Termas y Naumachias de los Céssares y Filipos en Tíboli y Antígola. Los Estanques, más benignos que los de Polión antiguos, más ferazes que los de Guadalope aora. Los reliebes vaxillos y sellados varros, retratos, luzernas, tritones, colosos, armas diferentes del Persa olvidado y del Indio descubierto, viendo aquí los ojos lo que solo oímos de la Roma subterránea en los Kyrques, Reynaudos, Claudios y Ricardos."

58. There are not many references to this French bookseller. In a letter sent by Martín Sarmiento to Juan de Yriarte, in December 30, 1743, we read:

"En los primeros veinte años de este siglo, no había librero alguno extraño en Madrid. Sólo vivía un Anison, descendiente de los Anisones franceses, que traía tal cual libro de fuera y le vendía según su antojo. Los demás eran libreros españoles, que entonces no se extendían más que a comprar y vender libros facultativos, que llaman de pane lucrando, v.g. de medicina, leyes y teología. Hoy han mudado de aspecto las cosas. No sólo los libreros hacen venir de fuera cualquier género de libros, a proporción del consumo, sino que también los libreros extraños, movidos del interés, nos inundan con repetidos catálogos de libros venales, convidándonos a que, con preferencia, se hagan venir de sus tiendas o almagacenes."

José Santos Prieto, *Reflexiones literarias para una Biblioteca Real*, Santiago de Compostela, Consello da Cultura Galega, 2002, 71. Jean Anison printed in spanish Félix Palacios *Curso chymico ... en el qual se enseña el modo de hazer las operaciones mas usuales en la medicina ... / Traducido del idioma frances ... por Don F. Palacios*, Madrid, J. Garcia Infançon para F. Anison, 1703.

59. In reference to Mateo de La Bastida, bookseller that had his library in the Mayor Street, "front of San Felipe". Just in this place, painter Velásquez exhibits his first portrait of Philip IV. Francisco Pacheco (1564-1654), *Arte de la pintura, su antiguedad y grandezas, descriuense los hombres eminentes que ha auido en ella... y enseña el modo de pintar todas las pinturas sagradas / por Francisco Pacheco,* Seuilla, Simón Fajardo, 1649; Bonaventura Bassegoda (ed), Madrid, Cátedra, 1999, 205: "(...) *habiendo acabado el retrato de su Majestad a caballo, imitando todo del natural, hasta el país, con su licencia y gusto se puso en la calle Mayor, enfrente de San Felipe, con admiración de toda la corte y envidia de los de l´arte, de que soy testigo".*

60. Alberto Montaner Frutos, "El político Don Fernando el Católico", in: Aurora Egido & María del Carmen Marín, *Baltasar Gracián. Estado de la cuestión y*

nuevas perspectivas, Zaragoza, Institución "Fernando el Católico", 2001, 49. Miguel Romera Navarro, "Antología Graciana", *Hispanic Review*, 14-1 (1936), 11-40.

61. José Enrique Laplana, J., "El Discreto", Don Fernando el Católico", in: Aurora Egido & María del Carmen Marín, *Baltasar Gracián*, 64.

CHAPTER SEVEN

THE SEVENTH DESK: MATHEMATICAL INSTRUMENTS, PHILOSOPHICAL ARTIFACTS AND THE SECRETS OF NATURE

MARÍA M. PORTUONDO

During the mid-seventeenth century, Vincencio Juan de Lastanosa (1607-1681) assembled in his hometown of Huesca a remarkable collection of curiosities, art and antiquities that he displayed in his home museum. Contemporaries considered him an expert in Iberian antiquities and an art collector with a fine taste. Hundreds of bronzes and marble statues graced the galleries of his home, part of his extensive collection of Iberian Roman antiquities. His most prized possession, and the sole aspect of his collection on which he published, was an extensive collection of Roman, Visigoth and Moorish coins and medals.[1] The items in the museum, however, were not limited to antiquities and artworks. Just as other continental collectors did during the seventeenth century, Lastanosa also possessed a large selection of *naturalia* and *artificialia.* Along with the curious or even bizarre works of nature that formed part of his *naturalia* collection, Lastanosa also displayed a number of exemplary works of human artifice. The collector clearly conceived of these artifacts as distinct from antiquities or works of art and displayed them in a deliberate and in a carefully staged manner. These works of artifice consisted of a number of mechanical contrivances, an armory, mirror displays, and a collection of what we would characterize today, anachronistically, as "scientific" instruments. An extensive library of natural philosophical and technical works, with what were surely some well-thumbed volumes, complemented the artifact displays.

This essay focuses on the instruments and the related collection of scientific books that Lastanosa kept in his museum in Huesca. My aim is to understand the representational and didactic roles these items played in Lastanosa's museum and approximate a characterization of Lastanosa's ideas about nature and science. I base this work on the premise that within the culture of early modern cabinets of curiosities artifact displays were understood by collector and visitor alike to inhabit a multi-dimensional symbolic space.[2] In other to reconstruct this space, we must unravel the system of correspondences that informed the viewer's understanding of the artifact, but we must refashion as well—even if only in our mind's eye—the material nature of the artifacts.

Reconstituting the material quality of Lastanosa's instrument collection is, unfortunately, the most difficult aspect of this exercise.[3] The artifacts no longer exist. By the beginning of the 18th century, Lastanosa's collection had largely vanished, having been sold off, pilfered or otherwise scattered. What we do have, thanks to the efforts of the Lastanosa Project under the sponsorship of the Instituto de Estudios Altoaragoneses, are a number of authenticated first-hand descriptions of the collection, including one written by the collector himself. The Project has located descriptions and inventories at the Royal Library in Stockholm[4] and at the Hispanic Society in New York[5] and throughout Spain. These, as well as descriptions made by Lastanosa's contemporary and friend Juan Francisco Andrés de Uztarroz, provide rich sources from which to reconstruct the collection of instruments and books Lastanosa amassed.[6]

A Home as Mirror of the World

Before we examine in detail Lastanosa's collection of instruments, we should first visit the rest of this remarkable house. The statue of Hercules holding the celestial orb that crowned the tower on the front of the house surely suggested to the unsuspecting visitor that one was about to enter a most uncommon place. Lastanosa's collection was arranged throughout the public rooms of his large house, which was also home to his family of fourteen children. The displays began on the first landing of the principal staircase that led to two floors of galleries. A smallish room off this main landing must have been the first stop during a visit to the museum. Lastanosa displayed in this room a number of artifacts that represented the "subtle fulfillment of mechanics as admirable efforts of artifice" (*admirables esfuerços del arte en sutilissimos desempeños de la Mecanica*).[7] A contemporary described it as the "famous chamber" or the "*célebre camarín.*"

A colorful glass window let light into the *camarín* which was decorated with mirrors on elaborate frames made of precious woods or silver. These shared the walls with canvas paintings and bas-reliefs fashioned from lead, gesso and cardboard. A convex mirror cast the view of the whole room "as in perspective," while a concave mirror created the illusion of suspending in mid-air any object that fell within their range of reflection (*arroja fuera de si y representa en el ayre el objeto que se le acerca*).[8] It was room artfully designed to create illusions and play with the sense on vision.

Uztarroz and Lastanosa both explained that these mirror arrangements were inspired by those described in the French book *Mathematical Recreations* by Jean Leurechon.[9] Both writers, however, left unacknowledged the fact that these had originally appeared in Book XVII of Giambattista Della Porta's influential manifesto on the practices associated with natural magic, the *Magiae naturalis* (1558).[10] In this book, Della Porta explained that the practitioner of natural magic, once in possession of the secrets of nature, could use this knowledge to duplicate natural wonders. The magus accomplished this by unraveling the interrelation and sympathies that existed between the hidden forces that controlled nature. Della Porta's book was one of the first examples of a genre akin to today's works of popular science. These books taught readers how to duplicate the recipes, cures, mechanical contrivances and such. Moreover, they encouraged the "curious reader" to attempt new configurations and thus fashion new experiences for themselves. Lastanosa took this advice to heart.

Throughout the museum, the collector installed arresting optical displays designed to increase the visitors' sense of wonder. We cannot underestimate the novelty of this use of curved mirrors. During the early seventeenth century, fine glass mirrors remained impossibly expensive, difficult to manufacture and could only be procured from Venice.[11] It was only during the mid-century that mirror-making technology reached the point were it was possible to make concave mirrors capable of projecting images.[12] Leurechon describes them as made from globe sections and ranging from two to four inches and upwards of three feet in diameter. We do not know the size or curvature of Lastanosa's mirrors but he managed to project an image just as Leurechon described.

Leurechon explained a bit dramatically what happened when an unsuspecting observer encountered this effect, "Now the image coming thus to the Eye, those which know not the secret, draw their Sword when they see an Image this to issue out of the Glass, or a Pistol which some one holds behind."[13] When Lastanosa decorated the *camarín* with convex and concave mirrors, he was trying to replicate this sense of drama and surprise in his visitors. Yet, the *camarín* was also designed to send an additional message. The use of mirrors during the mid-seventeenth century also carried the implication that their owner understood the optical—and perhaps even magical—secrets of mirrors. The *camarín* served as the antechamber to the other marvels visitors encountered as they continued to the larger galleries.

Exempla from Leurechon's book did not end with the curved mirrors that hung on the walls of the *camarín*. Three large desks occupied the center of the *camarín* and each displayed objects considered the state of the art of artisanal production.[14] The first displayed a "triangulated obelisk" (*obelisco triangulado*) fashioned on a lathe. The term *trianglulado* likely refers to a wooden piece turned using an eccentric lathe, and represented the greatest achievement of what a woodcrafter could make with that tool. During the late sixteenth and seventeenth century, items produced by increasingly sophisticated and automated lathes were much sought after items in cabinets of curiosities.[15] An example of another mechanical wonder sat on the next table. It displayed a miniature representation of an infantry battle scene. Carefully placed mirrors gave the appearance that the small figurines dressed in military costumes and carrying tiny arms marched against each other as if engaged in battle. A spinet imitated the sound of the battle trumpet. The vignette appeared as the eighty-first problem of the *Mathematical Recreations*, as part of the author's explanation of the effect of placing flat mirrors opposite each other. Finally, the artifacts displayed on the third desk where a material representation of the wonders that resulted when the natural and the man-made intersected: the head of a Negro carved in jet (*azabache*) and carved boxes made from seashells.[16]

Other cabinets in the small room opened to reveal hundreds of compartments and drawers holding part of Lastanosa's collection of natural prodigies, a cornerstone of his collection. Uztarroz explains that the compartments held "the most diverse curiosities, both mechanical and admirable things of nature, of which a noteworthy treatise could be written of the prodigies enclosed within the earth's fruitful womb."[17] Next to this display there was a large bookcase with even more compartments and shelves, the third of which held books on mathematics, astrology, cosmography, fencing or "the art of the sword," chemistry and plants. Two volumes were particularly important and were singled out by Uztarroz; one was a papyrus scroll kept in an ivory box that contained a "Kabbal or art of divination" and the other a curiously folded prognostication kept in a silver box.[18]

When visitors left the *camarín*, they did not leave Lastanosa's fascination with mirrors behind. The larger galleries displaying artworks and antiquities were also decorated with mirrors. In some instances, these duplicated the image from a painting that hung nearby. In other cases, the mirror itself contributed to the artwork in the gallery. In a room that displayed his largest painted canvases, Lastanosa intercalated between the canvases large convex mirrors that reflected into the room a view of his garden and the nearby fields.[19]

Just as Lastanosa's selection of decorative items and paintings spoke to his artistic sensitivities, Uztarroz's verse reminds us that visitors understood this use of mirrors as a manifestation of the collector's artistic vision. Uztarroz rhapsodized,

En espejos se mira delineada	One can see delineated in mirrors
quanta curiosidad está cifrada	the curiosities that are encrypted
en laminas, en lienços i en ornatos,	on prints, canvases and decorations
que los cristales a la vista gratos	that the mirrors, so pleasant to the eye,
muestran el artificio en lo luciente,	reveal their artifice in the reflection,
i en lo breve trasladan claramente	and transform clearly the objects within
los objetos que alcançan en retratos.[20]	their reach into paintings.

Lastanosa's use of mirror to create illusions reached its apotheosis during the spectacle staged in Huesca in 1658 to commemorate the long-awaited birth of the heir to the Spanish throne, Felipe Prospero. On the façade of the house, and as part of a number of allegorical representations staged on the adjacent plaza, Lastanosa build a replica of Vulcan's cave. Viewers could make out within a fiery grotto a number of Cyclopes working on forges. Over two hundred mirrors of different shapes and sizes captured the light to give the impression of the sparks flying off the monster's forge.[21]

The collector carefully arranged his catoptric displays in order to play with the observer's senses and with the clear intention of creating in the viewer a sense of astonishment. We do not get the impression from the surviving descriptions that Lastanosa was attempting a carnivalesque display, or that his intent was to disorient or make a fool of his visitors. He never seems to lead the visitor too far from what was real, since it was precisely within what nature and man could create that he sought to find that which fascinated. The mirrors complemented and magnified the general impression Lastanosa sought to leave with his visitors after they had toured the rooms of his house: that of having visited a remarkable place, where even the most ordinary artifacts possessed a sometimes not readily apparent element of the wondrous and magnificent.

Lastanosa's contemporaries prominently featured mirrors in paintings; witness the works of Vermeer and *Las meninas* of Velázquez (1655).[22]

The catoptric displays throughout the museum in Huesca betray Lastanosa's desire to join the company of the artists his visitors admired. He wanted to be considered an artist, but not one who wielded a brush or a chisel, but one who constructed mirages using one of the technological advances of his time: lenses and curved mirrors. Lastanosa was suggesting to his visitors that he was an artificer of the senses and a possessor of secrets. His house and garden were his canvas; his mirrors were his brushes.

Mathematical and Philosophical Instruments

This was the home of an artist and artificer, that of a collector who proudly displayed the source of his knowledge: books. The *camarín* had cabinets full of books, as did other rooms throughout the house. But it was upon entering the library that visitors understood this was the collection of an aspiring savant. The walls were decorated with a number of maps, as well as perspective drawings of various cities.[23] The room also served as the antechamber to a gallery that displayed the largest part of the collection of *naturalia* and to the armory.

Along one of the side walls of the library, flanked by a desk holding history books and by another with Lastanosa's impressive collection of antique coins, was the desk displaying his collection of instruments and philosophical artifacts.[24] This seventh desk held instruments used in geometry, mathematics, astrology, catoptrics, fortification and perspective.[25] This collection of mathematical instruments was comprehensive and exquisite. It is clear from their descriptions that both Lastanosa and Uztarroz valued these artifacts for the computational operations they performed, but they also considered them precious artifacts that represented the possibilities that lay at the intersection of craftsmanship and mathematics.

Anyone approaching the seventh desk recognized these as being the tools of a practical polymath. One who, if he mastered the use of these instruments, could survey the land and depict it on a map. One who could calculate the location of the stars and perhaps prepare an astronomical prognostication. One who could unravel the secrets of drawing using perspective. One who could design a fort, form a squadron of soldiers and determine the final resting place of a canon ball. One who could guide a ship at sea. One who knew about and used the latest optical instruments: microscopes and telescopes.

For geometrical calculations, Lastanosa had a number of excellent compasses. He was proud of the *pantómetra* (proportion compasse or sector). Lastanosa's sector was made in Amsterdam and was described as having "many uses," meaning it could be used to solve a number of different mathematical problems.[26] The instrument had flattened, engraved legs showing graduated scales used to solve a number of different mathematical problems. These instruments were perhaps only second to the astrolabe as the most sophisticated computational device of the time. The clever instrument facilitated arithmetic computations by using the geometric proportions embodied in the triangle. They were used generally to solve gunnery or architectural problems, although they were quite versatile. For example, the Jesuit mathematician José de Zaragoza (1627-1679) made a logarithmic pantometer that solved both military and musical problems as a gift to Charles II.[27] Lastanosa's pantometer dated from an earlier period but it is reasonable to suppose he would have also sought this example of mathematical craftsmanship from the Jesuit mathematician with whom he exchanged correspondence.[28]

For drafting, Lastanosa had bronze dividers with steel points also made in Amsterdam. Completing his collection was another bronze compass with interchangeable tips fitted with a semicircular sector bearing the degree scale and set to a particular angle using a screw. Complementing these calculation instruments was a fine selection of drafting tools: scale rulers, drafting triangles, and protractors. These were joined by a "bronze quadrant that folds very curiously."[29]

For astronomy and astrology, Lastanosa had quadrants, astronomical rings and astrolabes. We do not have specific descriptions of the astronomical rings or of the astrolabe, but the account describes a bronze quadrant, "with an excellent graduated scale on one side and on the other a sundial [*celindro*]." He also displayed an instrument made of walnut to determine elevations, level planes, take declinations and layout sundials. There were also a number of sundials, as well as magnetic compasses and loadstones.[30] The *Sparvenfeldt* inventory also mentions other noteworthy astronomical instruments: a large wooden sphere, a smaller one made of silver, and two sets of celestial and terrestrial globes, the largest one three hands in diameter, beautifully illustrated, "modern" with bronze meridians.

When gazing upon this collection from our modern perspective, we might hastily characterize the artifacts on the seventh desk as scientific instruments. Yet, this does not accurately describe all the instruments Lastanosa displayed. They belong to two separate categories: mathematical instruments and philosophical instruments or artifacts. There is a fine but important distinction between them. Mathematical instruments, such as astrolabes, rulers and quadrants, measure that which can be perceived readily by the senses. Philosophical instruments, such as telescopes, microscopes, on the other hand "see" beyond what the senses allow. These might also facilitate the construction of artificial or non-naturally occurring events and environments. Associated with these were other philosophical artifacts, such as prisms, curved mirrors and magnets.[31] James A. Bennett, the distinguished Keeper of the Museum of the History of Science at Oxford University, pointed out the conceptual divide that existed between mathematical instruments and philosophical instruments during the seventeenth century. As he explained, "insights into the natural world, its unseen structures and organizing principles, its causal links, its formal and material constitution, belonged with natural philosophy, whereas instruments belonged with mathematics."[32]

Lastanosa's fine pantometers, astrolabes and such—surely practical and intended for use—were tools used for measuring. Its user could quantify the unruly reality of the earth and heavens and translate observations into dimensions that could be manipulated, with mathematics and geometry acting as the means of translation. The seventh desk, however, also displayed philosophical instruments. Many of these owed their genesis to the tradition of natural magic and during the seventeenth century became a cornerstone of the new experimental sciences by broadening the range of inquiry into nature beyond that allowed by mathematical instruments.[33] Among philosophical instruments, Lastanosa had prisms, cylinders and mirrors (concave, convex, and hyperbolic), in addition to a collection of dioptric and catoptric instruments. For example, he had a pair of *anteojos* or magnifying eyeglasses that let the user "see a horrible monster in a tick" and "from tiny seeds see large fruits."[34]

Lastanosa shared with his contemporaries the enthusiasm for the telescope and understood it to be one of the wonders of his era.[35] Lastanosa had two "very good" telescopes. We know that one of them was 10 hands in length and had sufficient magnification to see sunspots and *Venus cornuta* or the Venus crescent.[36] He was familiar with the "new wonders" discovered on the Sun with Kircher's helioscope and on the Moon with Antonio Schyrle of Rheita's binocular telescope.[37] Schyrle mystical brand of natural philosophy, although anti-Copernican, openly speculated on the possibility of extraterrestrial life. Lastanosa might have shared this opinion with Schyrle, whose map of the moon he had surely studied.

Optical artifacts for Lastanosa functioned as gateways to previously unseen worlds and as devices through which one could witness strange marvels. During the seventeenth century, the telescope was thought to have been divinely revealed so mankind could further his inquiry into nature.[38] Therefore, when Lastanosa displayed publicly a microscope or a telescope, he understood these as instruments that opened new vistas into nature that only a privileged few had seen. He was also informing the visitor that its owner was a member of a community of savants, an adept in the secrets of nature and well informed of the latest discoveries.[39] He argued that such were the marvels these instruments uncovered, that if these discoveries had taken place during a past era when wise men defended the notion that the stars were indeed like the earth, they would have found their opinions confirmed by these observations. For with the telescope they would have seen clearly mountains, valleys, large provinces, and impenetrable seas with islands.[40]

The culture of curiosity also fostered the belief that many secrets of nature lay hidden in plain sight, waiting for a gifted mind to decode them. This was the foundational belief behind the branch of astrology that was of interest to Lastanosa. It posited that the heavens—the position of the stars and their relationship to the earth—could be read as if a book. It also suggested that ancient civilizations had understood this, had decoded this "language" and at one time practiced an astrology different from the Ptolemaic astrology prevalent in Europe. During the seventeenth century, deciphering the celestial orb also implied reading the heavens figuratively, in addition to determining astral influences based on celestial positions. This branch if astrology clearly appealed to Lastanosa's antiquarian interests.

As a conscientious collector, he had books representing many different branches of astrology, but he was most intrigued by the work of Athanasius Kircher and Jacques Gaffarel.[41] These authors asserted that the key to astrological interpretation of the heavens lay in understanding ancient languages. Both authors, one using Egyptian and the other Hebrew, had found that celestial constellations corresponded to characters in these ancient languages and thus presented a new way of preparing horoscopes. In Lastanosa's words, they had "put into practice the method of reading the extensive celestial book though knowledge of these characters."[42]

Lastanosa's interest in optics extended beyond creating optical illusions or seeing what formerly had been hidden, however. His "hyperbolic eyepiece" also served as a religious, contemplative aide. The eyepieces, he explains, "unite various fragments of divers objects (*en planos*), and form with perfection that which is not, representing the most high mysteries of our faith, and there is so much and the matter so extensive that a man's life is short to see and enjoy its operation."[43] For whereas during the Renaissance the act of gazing in contemplation into a mirror was framed as an exercise in self-examination—a way of examining one's soul—the devices of the seventeenth century built with curved mirrors and lenses fabricated illusions and images that were open to interpretation and public study.[44] We see in Lastanosa's case that this exercise could take place still very much within the Christian interpretive framework. A fine example of this is the work of Schyrle de Rheita, who claimed to have seen clearly represented through his binocular telescope the *sudarium* of Veronica in the sign of Leo.[45]

What then does Lastanosa's collection of philosophical artifacts tell us about him? Lastanosa belonged to a generation of intellectuals that was far from homogeneous in their approach to the exploration of nature. This generation negotiated the transition from Scholasticism and the sophisticated empiricism of the late Renaissance to the new science of Descartes, Huygens and Newton. It was likewise a time when the magic-mystical approach inherited from Della Porta coexisted with skepticism and the new experimental philosophy advocated by Francis Bacon. These different approaches to the study of nature often intersected and informed each other. In fact, Della Porta's *Magiae naturalis* described many predecessors to the instruments that in the hands of Galileo and Newton became the tools that yielded the modern experimental method and the construction of a new body of natural philosophical ideas based on facts.[46] A seventeenth century eclectic savant, such as Lastanosa, drew equally from the observational work of a Galileo or from someone, like Kircher, who approached the inquiry into nature from the natural magic perspective.

Lastanosa's description of his collection of mathematical and philosophical instruments reveals that he inhabited an intellectual space where curiosity and a sense of wonder guided the inquiry into nature; a space rich with symbolism, where invisible forces established a system of correspondences and interconnectedness that governed the operation of the natural world. It is also evident from his repeated references to the new discoveries made possible by optical devices that he conceived of optics as having potential beyond simply casting curious images and creating artistic illusions. For, as Leurechon reminded his readers, experiments with mirrors also opened the doors to the search for truth. The individual who approached mirrors and lenses with curiosity, the Jesuit counseled, "will have reflection of quantities of Truth, and fine secrets of Nature."[47]

Lastanosa mentions that he copied his setups from books written for a sophisticated audience of collectors, such as the works of Leurechon, Kircher and Bettini. Notably absent from his library are the works on optics by Descartes and Kepler, both far more mathematical and experimental than the books Lastanosa cites. Lastanosa undoubtedly carried out many trials until his mirror configurations worked to his satisfaction and patiently observed through telescopes until he witnessed the Venus crescent. The collector from Huesca indeed "experimented" with mirrors and lenses but not in the modern sense of the word, which implies a systematic and quantified investigation aimed at understanding natural phenomena.[48] He did so instead as a way of experiencing nature in a new and exciting way. For Lastanosa, the goal of investigation into nature was to witness new wonders and to share with others what he had uncovered, rather than to explain the causes behind the phenomena.

Library of Scientific Books

Lastanosa's fascinating display of art and artifacts was but a fraction of the items he collected. Another of his passions was amassing an impressive number of books, all carefully selected to complement his collection. The passage of time has been kind to the historian, for we have not just one catalog of books—the *Sparvenfeldt Catalog*—but also another text—the *Narración de lo que le pasó a Don Vincencio Lastanosa...con un religioso docto y grave*—that consists of Lastanosa's own observations about his book collection. In this text, under the trope of taking a visitor for a brief tour of his museum and library, Lastanosa lingers occasionally to point out particularly significant tomes and highlights those he found noteworthy. It is a narrated tour of those aspects of his intellectual universe he chose to make public. The précis of Lastanosa's collection of scientific books that follows takes into consideration the books in the *Sparvenfeldt Catalog*, as well as those Lastanosa singled out in the *Narración*.[49]

Lastanosa grouped his collection of scientific books and manuscripts into the categories of: "mathematics, geometry, astrology, cosmography, hydrography, geography, perspective, optics, dioptrics, catoptrics, painting, architecture, military arts, timekeeping, use of weapons, fortification, troop formations, horse training, music and dividing the land [cartography]."[50] He appends the following disciplines to this list once he starts describing his collection: mechanical arts, natural philosophy, natural history, agriculture and gardening, animals, minerals, secrets and magic, medicine, surgery and alchemy.

On astrology, Lastanosa had the classic work of Ptolemy and the late sixteenth century work of F. Junctinus, as well as more current works by Andrea Argoli and perhaps Luca Gaurico's *Tractatus astrologicus*. He had the *Ephemerides* of Luis Freire based on Tychonic and Copernican cosmologies,[51] and the earlier Ptolemaic but revised Alphonsine Tables published in 1521 by the professor from Salamanca, Pedro Ciruelo[52] and in 1641 by García Ventanas.[53] The catalog also lists Antonio de Obregón y Ceres's *Discursos sobre el astrologia moral* (Valladolid, 1603) and Antonio de Nagera, *Suma Astrolojica* (Lisbon, 1632).

He had the very best books on cosmography and cartography. He had what the catalog describes as a folio edition of Ptolemy with maps apparently published in León in 1535, likely the edition of Miguel Servet[54] From sixteenth century cosmographers, he had books by Sebastian Münster and the works of Peter Apian/Gemma Frisius in a Spanish edition of 1548.[55] He complemented these with several manuscripts on the theory of sphere. On cartography specifically, he had of the master Gerard Mercator, the *Atlas Minor* and from his disciple, Abraham Ortelius, his *Thesaurus geographicus* (Antwerp, 1596), the *Theatrum Orbis Terrarum* (Antwerp, 1602) and the *Epitome theatri Orteliani* (Antwerp, 1601).

He does not single out the books on descriptive geography in the *Narración*, but he had an impressive collection, beginning with a Spanish edition of Pomponius Mela's *Compendio geografico del Orbe* (Madrid, 1644). He had forty nine matching volumes of Elzevir published collection of republics and complemented these with Philipp Clüver's introduction to ancient and modern geography, *Introductionis in universam geographiam* (Leiden, 1642). He had both a manuscript and a printed copy of Joan Botero Benes, *Relaciones universales del mundo* (Valladolid, 1603). Two geographical works with regional interests complemented the universal cosmographies: José Sesse y Piñol's *Libro de la cosmographia universal del mundo y particular descripcion de la Syria y Tierra Santa* (Zaragoza, 1629) and Tommaso Porcacchi's *L'Isole piu famose del Mondo* (Venice, 1604), an atlas of the principal islands of the world. Lastanosa's interest in understanding the world even extended to imaginary worlds. Gracing his shelves were editions of Thomas Moore's, *Utopia* and Hall's satirical critique, *Mundus alter et idem.*

Closely associated with cosmography was the art of timekeeping or horology. Lastanosa prized a manuscript book in nine treatises or chapters on sundials written for him in 1627 by Padre Jaime Alberto, a Valencian Jesuit, whom he describes as beloved friend (*amantissimo amigo mio, escrito para mi enseñanza, adonde ay artificiosos y primorosos reloxes*).[56] His catalog also lists Pedro Roiz's *Reloxes solares* (Valencia, 1575). Lastanosa also had books on how to make all manner of sundials by Pierre de Floutrières, (Paris, 1638) and several editions of the Jesuit Pierre Bobynet's book on gnomics.[57] From his collection of Kircher's books, he pointed out the *Ars magnetica* and *Ars lucis et umbrae*. The ninth book on gnomics in the *Apiaria universae philosophiae mathematicae* (Bologna, 1642) from another Jesuit, Mario Bettini, would have served Lastanosa as a helpful guide to build the sundial in his garden. Bettini's intriguing book also contained sections on perpetual motion machines and curious mechanical contrivances.

On optics, he mentioned the Jesuit François de Aguilón's book, probably his *Opticorum* (Antwerp, 1613). On dioptrics, in addition to the books by Schyrleus de Rheita, Kircher and Leucheron, Lastanosa highlighted Bettini's *Apiaria,* likely referring to books six and seven of the first volume, the *Dioptrica arcana* and *Catoptrica paradoxa*. Lastanosa noted that the literature on optics is so "perfected" that there were even books written on how best to polish mirrors and lenses, as Emmanuel Maignan did in his *Perspectiva horaria* (Rome, 1648).

On land surveying, he had the 1554 Paris edition of Roman surveyors Siculus Flaccus, Sextua Julius Frontinus, Agennius Urbicus and Hyginus Gromaticus.[58] For a contemporary manual on the subject, he mentioned his edition of Juan Perez de Moya's multi-volume *Tratado de matemáticas* (Alcala, 1573). While on the topics of navigation and hydrography, he proudly made reference to the work and achievements of another native son of the Altoaragon, Pedro Porter y Casanate. The native of Zaragoza had a distinguished military and naval career, serving as captain general of the province of Chile. He undertook a hydrographic reconnaissance of the coast of California seeking suitable ports to initiate the colonization of the area (1644-1652). Early in his career, Porter y Casanate wrote a critique concerning common errors in navigation *Reparo a errores de la navegación española* (Zaragoza, 1634).[59]

On the science of drawing using perspective, he highlighted owning the books by Vignola (Giacomo Barozzi), Albrech Dürer on perspective and symmetry,[60] and Daniel Barbaro's edition of Vitruvius.[61] Among those he considered modern authors, he had the three-volume work *La perspective practique* (1642) by "the Parisian Jesuit" (Jean Dubreuil), as well as those by Abraham Bosse on perspective and engraving, Jacques Aleaume,[62] and the *Perspectivae sintagma* of Hans Vredeman de Vries (Antwerp, 1626).[63] His catalog also included Girard Desargues's *Perspective* (Paris, 1648) on how to draw using perspective and the geometry of scales.

On the theory of modern and classical architecture, he had the requisites: Vitruvius,[64] Serlio,[65] Cataneo,[66] Jan Vredeman de Vries, and Roland Freart's *Parallel of Ancient and Modern Architecture*.[67] Supplementing these were volumes specializing in military architecture: Adam Freitag,[68] Nicolas Goldman,[69] Szymon Starowolski, as well as a book on cavalry drills and tactics by Lodovico Melzo and Flavius Vegetius Renatus's, *Epitoma rei militaris.* Mechanical wonders and military machines were represented by Augostino Rameli's book of military machines.[70] Such a collection would have not been complete without Diego Beson's (or Jacques Besson), *Theatro de instrumentos mathematicos* (Lyon, 1602).[71]

Lastanosa's book collection reflected his past as a captain during the war against the French in Catalonia (c. 1642). He complemented his books on military fortifications with those on the art of the sword with books by Achille Marozzo (Modena, 1536), Ridolfo Capoferro (Siena, 1610) and Luis Pacheco de Narvaez's *Compendio de la filosofia de las armas*, (Madrid, 1626). Crowning the collection on fencing was the majestic Elzevier folio edition of Girard Thibault's, *Academie de l'espée* (1626). On troop formations, he had printed titles by Miguel Pérez de Xea (Madrid, 1632), Miguel Lorente Bravo (Zaragoza, 1644), Lelio Brancacho (Barcelona, 1639) and Bernardino de Escalante (Brussels, 1595), in addition to a number of manuscripts.

Lastanosa's inclusion of the military arts, as well as fortification, troop formations, horse training into his taxonomy of the mathematical sciences attests to the increasing fusion of science and technology during the early modern era. There was a concentrated effort to mathematize and explain the physical phenomenon behind what had previously been the sole domain of the master craftsman.[72] During the sixteenth century, Spain led Europe in producing the most sophisticated of these works, principally in the art of navigation and artillery.[73] These extraordinary books attempted to codify the precepts of the military arts by subsuming it under the model of the sciences. They integrated mathematics and geometry into aspects as disparate as calculating the optimum number of soldiers needed to form a symmetrical squadron, the angle and trajectory of a cannon ball and the ideal marching cadence of soldiers goingto battle.

Although Lastanosa did not single out specific mathematical books or books on instrument making in his *Narración*, his collection included two important books on this discipline written in Spanish authors: Juan Pérez de Moya's *Aritmética* (Madrid, 1624) and Rodrigo Zamorano's first Spanish translation of Euclid's *Geometry* published in Seville in 1576. He also had a number of manuscripts that described how to make and use mathematical instruments. He mentioned one on the Jacob's staff (*vaculo astronómico*) and other manuscripts that explained how to use the pantometers and other mathematical instruments he displayed on the seventh desk.[74]

Lastanosa listed under the rubric of "Natural Philosophy" a group of works that more appropriately would fall under the category of natural history. On his shelves were the requisite Pliny,[75] Dioscorides in Andrés de Laguna's commentary and José de Acosta's *Natural History of the Indies* (Seville, 1590 and Barcelona, 1591). Gracing his shelves were other now-classic sixteenth and seventeenth century works on the discipline, such as Pietro Andrea Mattioli and Leonard Fuchs's large folio with painted prints, (Basel, 1542). His selection of books depicting animals was just as impressive. He had Conrad Gesner's *Historiæ animalium* in the lavish 1558 Zurich folio edition in color. On birds he had, Diego de Funes y Mendoza. *Historia general de las aves y animales* (Valencia, 1621) and Francisco Marcuello, *Historia natural de las aves* (Zaragoza, 1617). On domesticated birds, he had Juan Bautista Xamarro, *Conocimiento de las diez aves menores de jaula* (Madrid, 1604) and Jean de Franchières book on the breeding and training of hawks. Other authors in this volume discuss his extensive collection of medical, botanical and alchemical books.

Book and Instrument

After visiting the *camarín* with its mirrors and curiosities, after having stopped to gaze at Lastanosa's mathematical instruments and philosophical artifacts and after reviewing his library of scientific books, we can return to the question that motivated this essay. What does this collection tell us about Lastanosa's intellectual world? I believe there are three principal aspects of Lastanosa's scientific thinking that his library and collection of mathematical and philosophical artifacts bring to light.

First, it readily becomes apparent after thumbing through copies of the books Lastanosa signaled out in the *Narración* that the collector from Huesca favored practical books; books that taught him how to apply new techniques, how to use instruments, how to read the heavens and how to manipulate artifacts to achieve awe-inspiring effects. Using modern, although somewhat anachronistic terms, we might say that Lastanosa preferred application-oriented works to theoretical books. It might be more accurate to borrow the words of J. A. Bennett to say that Lastanosa preferred books that taught him how to "do," to books that explained what he was supposed to "know." Lastanosa sought a result or effect from his interests and pursuits, whether alchemical trials, painting with mirrors and lenses, or cultivating exotic plants in his garden. He was not a passive participant (consumer?) of early modern science but rather one who engaged in and fully experienced the wonders and marvels his instruments and artifacts revealed. His careful selection of instruments and related books suggests he engaged the scientific literature of his time and was current on the scientific debates of his era.

The second aspect I would like to emphasize is Lastanosa's preference for the work of Jesuit natural philosophers; he highlights in the *Narración* at least seven books by Jesuit natural philosophers and mathematicians. He owned all of Athanasius Kircher's books and his letters to him suggest he held him in the highest esteem.[76] His contact and friendship with Jesuits residing in Huesca and during his visits to Madrid are a well-know aspect of Lastanosa's intellectual milieu.[77] He was the principal patron of Baltasar Gracián, another Jesuit and one of the most influential Spanish essayists of the time.

We lack a clear statement from our erudite collector about his natural philosophical ideas, but I believe Lastanosa's lay within the borders of seventeenth century "Jesuit science."[78] Recent scholarship has show that Jesuit science was not monolithic, and although committed to Aristotelianism, had sites such as Kircher's museum at the Roman College where diverging views about the natural world were negotiated albeit as an attempt to reconcile novel ideas with Catholicism. Jesuits interested in science faced significant restraints from their superiors as to how far they could deviate from Peripatetic orthodoxy, but managed to circumvent censorship by writing erudite and even obscure works that presented novel scientific postulates.[79] Within what Víctor Navarro has described as Jesuit eclecticism,[80] a restless intellect like Lastanosa's found both the comfort of religious orthodoxy and the excitement of scientific inquiry.

It surely appealed to Lastanosa that the Jesuits favored a culture of representation that encouraged demonstrations and spectacle with the dual aims of proselytizing and highlighting Christian learning at the service of God. This culture found its maximum expression in Kircher's museum at the Roman College. Lastanosa must have recognized in Kircher a kindred spirit: another keeper of many secrets, an artificer of the senses and an encyclopedic collector. However, Jesuit scientists followed a different episteme than the experimental philosophy that sought certainty in experimental results. Jesuit scientific practice found its proofs by reconciling Aristotelian natural philosophy with demonstrations. By the end of the seventeenth century, this type of scientific practice and its reliance on spectacle lost prestige within the scientific community, especially in England.[81]

Finally, I would like to return to the seventh desk and see if it can tell us anything more about the man who assembled the collection. Lastanosa's personal account of the contents of the seventh desk mentions that the artifacts on the desk were displayed accompanied by printed and manuscript books explaining how to use these devices. ([*Y*] *a todos los acompañan libros impressos o manuscritos que explican y enseñan su uso.*)[82] Uztarroz also emphasizes this mode of display and Lastanosa repeated the reference to book/artifacts displays when he discussed his collection of optical instruments.

Lastanosa, therefore, deliberately placed books next to instruments on the seventh desk. He did not, that we know of, place books on Roman antiquities next to his marble statues or books on perspective next to his Italian masters. What, then, did this particular display—instrument and book—imply? I believe this combination suggests that Lastanosa did not display the instruments and other scientific artifacts solely to be gazed upon because of their aesthetic merits or as examples of refined craftsmanship. Clearly they were exquisite, rare and surely expensive, two qualities that spoke to their owner's refined taste and deep pockets. Nevertheless, when considered relative to the rest of Lastanosa's collection, the seventh desk was but a fraction of the overall impression of grandeur that Lastanosa surely wanted to convey.

Lastanosa staged the display of mathematical and philosophical instruments side-by-side with books explaining their use to suggest something very specific to his visitor. I believe the combination book-artifact was intended to suggest that these instruments and artifacts were not passive entities, but that they instead embodied an active potential. In coming to this conclusion we have to understand that historians of seventeen century *wunderkammern* assure us that the instruments in these cabinets of curiosities did not see much, if any, use. As Derek de Solla Price explained, before the seventeenth century, instruments were meant to serve as representational models rather than as measurement instruments. The instrument embodied a theory; they were in fact called "theoriks." He explained that the principal function of pre-modern instruments was "that of an embodied theory, a jeweled ikon of the intellect, an altarpiece dedicated to understanding and entertainment."[83] Lastanosa's display of book and instruments was a reaction to this kind of passive and inherently contemplative attitude. What Lastanosa wanted to convey was precisely the opposite, that his mathematical instruments and philosophical artifacts *were used.* As the owner of both text and tool, Lastanosa was telling his visitors that he mastered the knowledge necessary to use what he so proudly displayed.

In the context of a cabinet of curiosities, philosophical instruments were understood as devices that allowed the user to encompass the natural world beyond what the senses could perceive unaided. While mathematical instruments operated to integrate into a value certain principles and models of the universe that we are otherwise unable to estimate. We cannot see the roundness of the heavens or know the location of the sun during the night, but the astrolabe can tell us. When latitude is calculated or the time of day determined using a sundial, the instrument performs an operation that reveals symmetries that nature hides and that only the instrument's user can uncover. Moreover, instruments can only reveal secrets if its owner knows how to use them and understands the principles behind them. As owner of book and instrument, Lastanosa was claiming that he was capable of revealing the hidden wonders of nature and of the man-made world. Revealing these required deciphering the use of an instrument to which Lastanosa, as owner and user, held the key. What better, yet discreetly veiled impression to leave in your visitors?

Notes

I would like to thank the Instituto de Estudios Altoaragoneses for their invitation to contribute to this project and for sponsoring such a worthwhile conference. Carlos Garcés Manau prepared an expert dossier of the latest scholarship on Lastanosa that was invaluable for this study. I specially want to thank the conference coordinators and editors of this volume, Mar Rey Bueno and Miguel López for their stewardship and collegiality.

1. Vincencio Juan de Lastanosa, *Museo de las medallas desconocidas españolas* (1645) and *Tratado de la moneda iaquesa* (Zaragoza, 1681).

2. Patrick Mauriès, *Cabinets of Curiosities* (London: Thames & Hudson, 2002), 25, 43, and Lorraine Daston and Katharine Park, *Wonders and the Order of Nature, 1150-1750* (New York: Zone Books, 1998), 91. See also the introduction to Oliver Impey and Arthur MacGregor, eds., *The Origins of Museums: The Cabinet of Curiosities in Sixteenth and Seventeenth-century Europe* (Oxford: Oxford University Press, 1985).

3. Recent historical scholarship on cabinets of curiosities and collections has overlooked Lastanosa's collection, a problem exacerbated by the fact that there has been virtually nothing written outside Spain about Lastanosa as collector. To make matters worse, what had been written was based on incorrect information. In 1912, the journal *Revue Hispanique* published an article by Adolphe Coster that reproduced a fictitious description of Lastanosa's collection found in a manuscript at the National Library in Madrid (Ms. 18727). On this subject, see Fermín Gil Encabo, "La ficción "telamoniana" de Pellicer en torno a Lastanosa", Actas del V Congreso de la Asociación Internacional Siglo de Oro Münster, 1999, Madrid, 2001, pág. 623-634; Fermín GIL Encabo, "Lastanosa y Gracián: en torno a Salastano", Actas I Congreso Internacional "Baltasar Gracián: pensamiento y erudición" (Huesca, 23-26 de mayo de 2001), Huesca-Zaragoza, 2003, 1:19-60 and Carlos Garcés Manau, "Lastanosa y la gran falsificación," Diario del Altoaragón, 20-Enero, 3 y 17-Febrero y 3 y 17-Marzo-2002.

4. The *Sparvenfeldt Catalog* at the Royal Library in Stockholm is a catalog of Lastanosa's extensive library of printed and manuscript books, including an inventory and description of his mathematical and philosophical instruments. Royal Library of Stockholm, Manuscript 379. *Sparvenfeldt Catalog of the Library of Vincencio Juan de Lastanosa.*
5. This document includes Lastanosa's description of his mathematical and philosophical instruments. Hispanic Society of America (New York), Manuscript B-2424, f. 52r-79v. Vincencio Juan de Lastanosa. *Narración de lo que le pasó a Don Vincencio Lastanosa a 15 de octubre del año 1662 con un religioso docto y grave.*
6. The description by Uztarroz of the *camarín* and the instruments is found in the *Descripción del Palacio y los jardines* and briefly in his published poem *Descripcion de las Antigueadades.* Hispanic Society of America (New York). Manuscript B-2424, f. 24r.-51v. Juan Francisco Andrés de Uztarroz, *Descripción del Palacio y los jardines de Vincencio Juan de Lastanosa.* Biblioteca del Palacio Real de Madrid, IX/5024 and the *Descripcion de las Antigueadades I jardines de Don Vincencio Juan de Lastanosa, hijo i ciudadano de Huesca, ciudad en el Reino de Aragón* (Zaragoza: Diego Dormer, 1647). Also valuable is the anonymous account of the festivities staged in 1658 to celebrate the birth of the heir to the Spanish throne. Biblioteca Nacional, VE 63-40. Anónimo. *Relacion de las fiestas que la ciudad de Huesca de el Reyno de Aragon ha hecho al nacimiento del Principe nuestro Señor D. Felipe Prospero* (1658).
7. *Relacion de las fiestas*, p. 6.
8. Uztarroz, *Descripción del Palacio*, f. 25r. It is unclear whether Lastanosa also replicated the famous *camara oscura* with a lens Della Porta described for the first time.
9. Leurechon, Jean. *Recreations mathématiques: composées de plusieurs problèmes plaisans & facetieux d'arithmetique, geometrie, astrologie, optique, perspectiue, mechanique, chymie, & d'autres rares & curieux secrets : plusieurs desquels n'ont iamais esté imprimez ... le tout representé par figures* (A Rouen: Chez Charles Osmont, 1629). The book is not in the *Sparvenfeldt* inventory but Uztarroz and Lastanosa refer to it. For a bibliographical study and the problematic attribution of the book to Jean Leurechon, see Albrecht Heeffer, *Récréations Mathématiques (1624). A Study on its Authorship, Sources and Influence* (October 7, 2004, cited April 12, 2007); available from http://logica.ugent.be/albrecht/thesis/Etten-intro.pdf.
10. The definitive study of Della Porta's *Magiae naturalis* remains, William Eamon, *Science and the Secrets of Nature: Books of Secrets in Medieval and Early Modern Culture* (Princeton, N.J.: Princeton University Press, 1994), 194-229.
11. The availability mirrors only increased to meet the high demand for the luxury item after France and later Germany and England began to compete with Venetian manufacturers. Sabine Melchior-Bonnet, *The Mirror: A History. Preface by Jean Delumeau*, trans. Katharine H. Jewett (New York: Routledge, 2001), 30-34.
12. Sara J. Schechner, "Between Knowing and Doing: Mirrors and Their Imperfections in the Renaissance," *Early Science and Medicine* 10, no. 2 (2005).

13. Jean Leurechon, *Mathematical recreations, or, A collection of many problems extracted out of the ancient and modern philosophers: as secrets and experiments in arithmetick, geometry, cosmography, horologiography, astronomy, navigation, musick, opticks, architecture, statick, mechanicks, chymistry, water-works, fire-works, &c* ... (London: Printed for William Leake and John Leake, 1674), 152.
14. Francois Filhol, a French priest and fellow collector with whom Lastanosa exchanged books and artifacts, had a similar display at his museum at the cathedral of Toulouse. Uztarroz describes the Frenchman's display of fine examples of craftsmanship as, *"No menos luce la curiosidad en las cosas antiguas que en las modernas, pues se hallan juntas en una estancia las obras i examnenes de diferentes maestros, quales son el de un tórnero, de un reloxero, que se labró en Estrasburch; de un platero, de un pespuntador, de un cerrajero, de un bordador, de un armero, de un escultor i de un pintor, cuyo lienzo fue presente de la Empreratriz."*
Filhol's collection also featured a significant and surely costly mirror display. Again Uztarroz: *"Assi mismo se hallan ... los instrumentos de la Geometria, Mathematica, Astrologia, Optica, Prespectiva, Catoptrica, i de la Dioptrica, con sesenta espejos de fundicion, con demostraciones varias, que dan mucho gusto i admiracion a los benemeritos de las artes liberales."* Juan Fracisco Andrés de Uztarroz, *Diseño la insigne i copiosa Bibliotheca de Francisco Filhol* (Huesca 1644). Published in, Ricardo del Arco y Garay, *La erudición española en el siglo XVII y el Cronista de Aragón Andrés de Uztárroz*, (Madrid, 1950) 2: 994, 995.
15. For the importance and preeminence of lathe-turned artifacts in seventeenth century cabinets of curiosities see, D. Graham Burnett, *Descartes and the Hyperbolic Quest: Lens Making Machines and their Significance in the Seventeenth Century* (Philadelphia, PA: American Philosophical Society, 2005), 25-26.
16. Historians Daston and Park have argued that these elaborate marriages between works of nature and the artisan's craft were an affront to the long held Aristotelian division between the natural and the artificial. The purpose of displaying these works was to "subvert the distinction by making art and nature indistinguishable." Daston and Park, *Wonders and the Order of Nature*, 277.
17. Uztarroz, *Descripción del Palacio*, f. 27r.
18. Uztarroz, *Descripción del Palacio*, f. 27v.
19. *"Ay assi mismo otros espejos mui crecidos, unos guarnecidos de evano y otors de plata, y en ellos, ya porla miltiplicacion, ya por la oposicion, se representan lo que se ven el los planos [en los lienzos] , y al contrario obran maravillosos efectos, que no solo recrean la vista sino que causan grande admiracion al entendimiento."* Uztarroz, *Descripción del Palacio*, f. 37r.
20. Uztarroz, *Descripcion de las Antigueadades*, 10-11.
21. *Relacion de las fiestas,* 14.
22. Melchior-Bonnet, *The Mirror*, 168-74 and Benjamin Goldberg, *The Mirror and Man* (Charlottesville: University Press of Virginia, 1985), 150-60.
23. Uztarroz, *Descripción del Palacio*, f. 40r.
24. The placement in the library of the fine collection Lastanosa and Uztarroz describe was not unusual; contemporary collectors also displayed their

mathematical instruments in their libraries. A. J. Turner, "From Mathematical Practice to the History of Collection," *Journal of the History of Collections* 7, no. 2 (1995): 140.

25. Lastanosa describes the desk as follows,
"Septimo escritorio. Contiene instrumentos para las Matematicas, como son esferas, anillos astronomicos, cilindrios, baculos mensorios, compases de proporcion, pantometras y otro gran numero, y a todos los acompañan libros impressos o manuscritos que explican y enseñan su uso. Ay un buen numero de cosas mecanicas de todos los artes y materias, espejos, antojos, concavos, convexos, hiperbolicos y otra mucha variedad, que por averse ablado de ellos arriba no se dice aqui con mas expression." Lastanosa, Narración, f. 72r-72v. Uztarroz's described the same desk as follows, *"El escritorio que se sigue está adornado varios cuerpos de Geometria, donde se depositan pantometras, compases de proporcion, quadrantes, anillos, astrolabios, niveles, esquadras, cartabones, vespertelio , perpendiculos , celindrios de la captotria y mathematicas, bruxulas, esteliones, piedra iman, calibros y muchos otros instrumentos para el uso de la Geometria, Mathematica, Astrologia, Captotria, Fortificacion, Prespectiva. Ay un libro en que están dibuxados la mayor parte destos instrumentos y sus maravillosas operaciones."*
Uztarroz, *Descripción del Palacio*, f. 40r-40v.

26. The sector was a seventeenth century refinement of the proportion compass. Michel Coignet, Galileo Galilei and Thomas Hood are credited with the simultaneous invention of the instrument. Erwin Tomach and Michael R. Williams, "The Sector: Its History, Scales and Uses," *Annals of the History of Computing* 25, no. 1 (2003).

27. Zaragoza describes the instrument's use in, *Fabrica y uso de varios instrumentos mathematicos* (1675). Cristina Bordas and Luis Robledo, "José Zaragoza's Box: Science and Music in Charles II's Spain," *Early Music* 26, no. 3 (1998).

28. Carlos Garcés Manau, "Un Lastanosa poco conocido (1665-1679). Las relaciones con Juan José de Austria," *Argensola* 115 (2005): 90. Félix Latassa summarized two no longer extant letters in Biblioteca Pública de Huesca, Mss. no. 76, *Memorias literarias de Aragón*, vol. 1, 29.

29. *Sparvenfeldt Catalog*, f. 100r.

30. Lastanosa also had the Jesuit N. Cabeus's book on magnets and early electrical experiments, although he does not mention this book in the *Narración* and it is not in the *Sparvenfeldt* Catalog. He refers to the book in a letter to Kircher, see note 77 below. Nicolaus Cabeus, *Philosophia magnetica, in qva magnetis natura penitus explicatur...* (Ferrara, 1629).

31. For more on this distinction, see Thomas L. Hankins and Robert J. Silverman, *Instruments and the Imagination* (Princeton, N.J.: Princeton University Press, 1995), 4-5 and A. J. Turner, "Interpreting the History of Scientific Instruments," in *Making Instruments Count: Essays on Historical Scientific Instruments Presented to Gerard L'Estrange Turner*, ed. R.G.W. Anderson, J.A. Bennett, and W.F. Ryan (Aldershot, Hampshire: Variorum, 1993), 25.

32. Jim Bennett, "Knowing and Doing in the Sixteenth Century: What Were Instruments for?" *British Journal for History of Science* 36, no. 2 (2003).
33. Thomas L. Hankins and Albert Van Helden, "Introduction: Instruments in the History of Science," *Osiris* 9 (1994): 3.
34. His description of the collection is worth reading in its entirety:
"Acompañan estos libros muchos instrumentos en que se ven raras maravillas. Para la Dioptica, antojos de los que aumentan los objetos, los crecen y los multiplican, algunos con notable admiracion, pues de una pulga se muestra un horrible monstruo, y de menudissimas semillas de flores crecidissimas frutas, como son cidras, limones, naranjas, melones y otra infinidad. De los antojos hiperbolicos, con que juntando varios fragmentos de diversos objetos en las tablas juntan y forman con perfeccion aquello que no ay, haciendo representar los mas altos misterios de nuestra fe, y es tanto y tan dilatado lo que ay de esto que es corta la vida de un hombre para mirar y gozar sus operaciones.
Para la Captotrica, espejos planos, convexos, concavos, cilindrios, hiperbolicos, con que se obran operaciones admirables, como son incendios, noches obscurissimas, inbiar luz a parte muy remota de modo que se vea y aun se pueda leer un papel. Los cylindrios recogiendo especies disipadas y dellas formando y representando figuras perfectas. Obran esto mismo los piramidales, concavos o convexos, los redondos puestos de lado, los convexos retirando en el centro el objeto representan en perspectiva toda una sala mostrando todos los lados de ella, de los concavos se ve la maior maravilla que es arrojar el objeto fuera y representarlo en el ayre. De los planos, que parecen los mas comunes y simples, opuestos unos contra otros se hacen las lontananzas, que dice el italiano, o una vista infinita que no tiene fin. Estos, multiplicados y rompidos los angulos multiplican el objeto, haciendo de un pedacillo de pais una distancia infinita, con tres o quatro hombres forman un exercito, y son tan universales sus operaciones de los unos y de los otros, que dello ay escritos muchos libros."
Lastanosa, *Narración*, f. 56r-56v.
35. Albert Van Helden, "Telescopes and Authority from Galileo to Cassini," *Osiris* 9 (1994).
36. "Algunos antojos de larga vista muy buenos, y entre ellos dos que tienen a diez palmos de largo y se ve con ellos las manchas del Sol, Venus cornuta, etc." *Sparvenfeldt Catalog,* f. 100v.
37. Lastanosa is referring to Kircher's map of the sun published in *Mundus Suberraneus, in XII Libros digestus* (Amersterdam, 1665) and to the work of Anton (or Antonius) Maria Schyrleus (also Schyrl, Schyrle) of Rheita (1597-1660) *Occulus Enoch et Eliae* (Antwerp, 1645) which include a map of the moon. For more on Schyrle, see Albert Van Helden, "The Telescope in the Seventeenth Century," *Isis* 65, no. 1 (1974): 44.
38. Antoni Malet, "Early Conceptualizations of the Telescope as an Optical Instrument," *Early Science and Medicine* 10, no. 2 (2005): 242-44.
39. Turner, "From Mathematical Practice," 140.
40. "*[S]iendo tales las maravillas que se descubren que si fuera en el siglo de aquellos sabios que defendieron ser otro cuerpo como el que nosotros habitamos, sin duda se confirmaran en esta opinion, pues con dicho instrumento se descubren*

distintamente montes, valles, dilatadas provincias, impenetrables mares con islas, peninsulas, con tan grande variedad y admiracion de los que las han visto que dellas han formado mapas, y aun distinguidolas con nombres." Lastanosa, *Narración*, f. 54v-55r

41. Athanasius Kircher, *Oedipus Aegyptiacus* (Rome, 1653) and Jacques Gaffarel, *Curiositez inouyes, sur la sculpture talismanique des Persans: Horoscope des patriarches. Et lecture des estoilles...* (Paris, 1631).

42. Lastanosa, *Narración*, f. 54vr.

43. Lastanosa. *Narración*, f. 55v-56r.

44. Melchior-Bonnet, *The Mirror*, 127-28.

45. Charles E. West, "The Binocular Microscope of the Seventeenth Century," *Proceedings of the American Society of Microscopists* 12 (1890): 58.

46. Hankins and Silverman, *Instruments and the Imagination*, 5

47. Leurechon, *Mathematical recreations*, 151.

48. Peter R. Dear, *Discipline & Experience: The Mathematical Way in the Scientific Revolution* (Chicago: University of Chicago Press, 1995), 11-31.

49. One historical problem remains unsolved, however. It is difficult to explain why many of the books mentioned by Lastanosa in the *Narración* do not appear in the *Sparvenfeldt Catalog* and vise versa. Ricardo del Arco y Garay wrote an annotated bibliography of the books Lastanosa singled out in the *Narración*. My book attributions differ in some instances from those made by Arco y Garay. This is likely the result of my use of the recent transcription of the *Narración* from the manuscript at the Hispanic Society and my reliance on the Union Catalog of all OCLC libraries (online at Worldcat) to ascertain bibliographical entries. Ricardo del Arco, *La erudición aragonesa del siglo XVII en torno a Lastanosa* (Madrid: 1934), 254-58, 345-49.

50. *"De las Matematicas, de la Geometria, de la Astrologia, Cosmografia, Hidrografia, Geografia, Perspectiva, Optica, Dioptica, Catoptica, Pintura, Arquitectura, Arte militar, de Reloxes, Destreza de las armas, de Fortificacion, de Formar esquadrones, de Adestrar cavallos, de Musica, de Dividir tierras..."* Lastanosa, *Narración*, f. 54r.

51. "664. Luis Freire de Silva, Cavallero Astronomo. Esfemerides generales de los movimientos de los cielos. En Barcelona año 1638. En 4º." *Sparvenfeldt Catalog*, f. 51r.

52. Manuscript described as: "2. Liber mathematicarum fragmentis. Astrologia coelestium rerum scientia. Petro Ciruelo, Joviano, Pontano, Josepho Scala et aliis nonullis, nullo servato rerum ordine delecta quaedam." *Sparvenfeldt Catalog*, f. 105r.

53. "331. Francisco Garcia Ventanas, Mathematico. Tabulae Alfonsinae perpetue motum coelestium. Matriti 1641. En 4º." *Sparvenfeldt Catalog*, f. 28v.

54. "185. Claudio Tolomeo. Su geografia y tablas. En folio grande. Leon 1535." *Sparvenfeldt Catalog*, f. 14r. Agustín Hernando identified the book as Servet's edition in "La colección de cartas geográficas reunida for Lastanosa (1607-1681)" *Argensola* 115 (2005): 121.

55. Peter Apian, Gemma Frisius, *Libro de la cosmographia: el qual trata la descripción del mundo y sus partes por muy claro y lindo artificio* (Antwerp, 1548). Listed in his catalog as item 63.
56. Lastanosa. *Narración,* f. 57v. For more on his relationship with Lastanosa, see José Ignacio Gómez Zorraquino, "El feliz encuentro de Vincencio Juan de Lastanosa y la Compañia de Jesús," *Argensola* 115 (2005): 105.
57. *Sparvenfeldt Catalog*, f. 50v.
58. *De agrorum conditionibus et constitutionibus limitum* (Paris: apud Adr. Turnebum, 1554.)
59. See introduction to, Pedro Porter y Casanate, *Reparo a errores de la navegación española [por] Pedro Porter y Casanate, y Navegación especulativa y práctica [por] José González Cabrera Bueno* (Madrid: Ediciones J. Porrúa Turanzas, 1970).
60. "11. Alberto Durero. De la simetria. Venecia 1591." *Sparvenfeldt Catalog,* f. 1v.
61. "214. Daniel Barbaro, Patriarca de Aquilea. Los diez libros de arquitectura de Vitrubio. Venecia 1584. 4°." *Sparvenfeldt Catalog,* f. 21v.
62. "666. La prespective speculative et practique per Aleaume, ingenieur du Roy. Paris 1643. En 4º." *Sparvenfeldt Catalog,* f. 51v
63. "835. Prespective Sintagma. Amstelodami 1626. En folio." *Sparvenfeldt Catalog,* f. 67v.
64. "644. Lucio Vitruvio. En ytaliano con comento. En folio. Como 1521." *Sparvenfeldt Catalog*, f. 50r
65. "890. Sebastiano Serlio. Sus libros de arquitectura. Paris 1545." *Sparvenfeldt Catalog*, f. 81r.
66. "826. Pedro Cataneo. Los 4 primeros libros de arquitectura. En folio. Venecia 1554." *Sparvenfeldt Catalog,* f. 67r.
67. "834. Paralele de la Arquitectura antique et moderne. En folio. Par Roland Freart, S. de Cambrai. A Paris MDCL." *Sparvenfeldt Catalog,* f. 67v.
68. "102. Adam Fritach. La arquitectura militar o la fortificacion nueba. En folio. En frances. A Leide 163568. *Sparvenfeldt Catalog,* f. 5r.
69. "661. La nouvelle fortification, de Nicolas Goldman. A Leide 1645. En folio. *Sparvenfeldt Catalog,* f. 51r.
70. "9. Agustino Rameli. Le diversi y artificiose maquine. Paris 1547." *Sparvenfeldt Catalog,* f. 1v.
71. The catalog entry likely refers to the Spanish edition of the popular "theater of machines" of Jacques Besson, *Teatro de los instrumentos y figuras matematicas y mecanicas ... / compuesto por Diego Besson ... con las interpretaciones de cada figura, echas por Francisco Beroaldo* (Lyon: Horacio Cardon, 1602).
72. Serafina Cuomo suggests that these types of military manuals, although sold as valuable in the battlefield, were aimed at "armchair warlords," who had the "capacity and inclination to learn from such books." She argues that they responded to a demand for political participation from educated members of society and for integration of military types into "normal" society. These texts often contained discussions about natural philosophy and "...contributed to the modification of some widely-held views not only about rest and motion, but also

about what form of physical phenomena, and how this knowledge should be used." Serafina Cuomo, "Niccolò Tartaglia, Mathematics, ballistics and the Power of Possession of Knowledge," *Endeavour: Review of the Progress of Science* 22 (1998): 31-35.

73. For a survey of Spanish contributions to navigation and the military arts, see José María López Piñero, *El arte de navegar en la España del Renacimiento*, 2 ed. (Barcelona: Editorial Labor, 1986) and Jorge Vigón, *Historia de la artillería española*, vol. 1 (Madrid: Instituto Jerónimo Zurita, 1947).

74. Lastanosa, *Narración*, f. 67v.

75. Among his editions of Pliny he had: Geronimo de Guarte's *Traducion de los libros de Cayo Plinio segundo de la Historia natural de los animales* (Madrid, 1599) and his *Traducion de la Historia natural de los pescados* (Madrid, 1603), Geronimo de Huerta's *Traducion de Cayo Plinio segundo* (First part in folio, Madrid, 1624) and Cristoforo Landino's Venetian edition of 1534.

76. Lastanosa explained to Kircher in a letter when he tried to purchase the Jesuit's works,

> "*Siendo aficionado a las buenas letras he procurado juntar una libreria adornada de antiguedades i instrumentos mathematicos, y deseandole dar su maior lustre deseo enriquezerla con sus obras de Vuestra Paternidad Reverenda, porque aunque tengo de antiguedades todo lo curioso i raro que ha salido de vente años a esta parte, i de las mathematicas tengo algunos modernos mui curiosos como son la Apiaria del Padre Mario Betino Bononiense Soc. Iesu en dos tomos, del Padre R. F. Antonio Maria Schirleo de Rheita Oculus Enoc et Eliae, del Jesuita Parisiense tres tomos en quarto de prespectiva, Sintagmata prespectiva, Cabeus De Magnete, Aquilon De Optica, y desta calidad de libros todos los modernos que han llegado a mi noticia, i a estos acompañan los del arte militar, arquitectura, fortificacion, manexo de las armas, cavallos y de la espada, y de los antiguos tengo todos los que se hallan destas mismas sciencias y artes, pero todo lo juzgo por poco careciendo de lo mas precioso que son los escritos con que Vuestra Paternidad Reverenda ha enrriquezido el mundo.*"

APUG 568, f. 187r-187v. Letter by Vincencio Juan de Lastanosa to Athanasius Kircher. Huesca, May 15, 1656. Published in, Carlos Garcés Manau, "Diez cartas de Vincencio Juan de Lastanosa y Diego Vincencio Vidania a Athanasius Kircher, conservadas en la Universidad Pontificia Gregoriana de Roma," *Argensola* 115 (2005): 187-199.

77. José Ignacio Gómez Zorraquino, "El feliz encuentro de Vincencio Juan de Lastanosa y la Compañia de Jesús," *Argensola* 115 (2005): 95-114. For more on the influence of Jesuits in Spanish science, see Thomas F. Glick, "On the Influence of Kircher in Spain," *Isis* 62, no. 3 (1971).

78. Some recent scholarship on Jesuit science, includes Peter R. Dear, *Discipline & Experience*; Mordechai Feingold, ed., *Jesuit Science and the Republic of Letters* (Cambridge, Mass.: MIT Press, 2003); Paula Findlen, ed., *Athanasius Kircher: The Last Man Who Knew Everything* (New York: Routledge, 2004); John W. O'Malley

et al., ed., *The Jesuits: Cultures, Sciences, and the Arts, 1540-1773* (Toronto: University of Toronto Press, 1999).
79. Mordechai Feingold, "Jesuit: Savants," in *Jesuit Science and the Republic of Letters*, 1-45.
80. Víctor Navarro, "Tradition and Scientific Change in Early Modern Spain: The Role of the Jesuits," in *Jesuit Science and the Republic of Letters*, ed. Mordechai Feingold (Cambridge, Mass.: MIT Press, 2003), 332.
81. Findlen, ed., *Athanasius Kircher*, 37 and Michael John Gorman, "From 'The Eyes of All' to 'Usefull Quarries in Philosophy and Good Literature': Consuming Jesuit Science, 1600-1665," in *The Jesuits: Cultures, Sciences, and the Arts, 1540-1773*, ed. John W. O'Malley et al. (Toronto: University of Toronto Press, 1999), 183-84.
82. Lastanosa, *Narración*, f. 72v.
83. Derek de Solla Price, "Philosophical Mechanism and Mechanical Philosophy," *Istituto e museo di storia della scienza* 5, no. 1 (1980): 76-78.

CHAPTER EIGHT

APPEARANCE, ARTIFICE, AND REALITY: COLLECTING SECRETS IN A COURTLY CULTURE

WILLIAM EAMON

In October 1662, a "learned and serious cleric" supposedly arrived at the home of Don Vincencio Juan de Lastanosa in Huesca to visit the collector's famous library and museum.[1] Like many others who visited Lastanosa, the cleric had come "from a faraway province" to see the remarkable collection of books, antiquities, and curiosities that Lastanosa had assembled over a period of more than twenty years. By 1662, Lastanosa's collection had become famous throughout the European world, attracting visitors and travelers from all over Spain and Europe. Huesca was a little off the beaten track in seventeenth-century Spain. Neverthless, Lastanosa's celebrated collection could boast visitors such as the dukes of Orleans, Ferrara, and Mirandola, the Prince of Squilache in Italy, and a host of Spanish notables.[2] King Philip IV, on his way from Zaragoza to Madrid, is even said to have paid the virtuoso a visit.

Lastanosa wrote an account of the cleric's visit in an unpublished work titled *Narración de lo que pasó a Don Vincencio Lastanosa con un religioso doctor y grave*. It is probable, as Ricardo del Arco suggests, that Lastanosa's account of the visit was imaginary – that is, that the *Narración* was a fiction.[3] In fact, Lastanosa himself gives away the conceit when, in the dialogue reconstructing the visit, he informs the cleric that, since his health is poor, he will have to guide the tour from his chair: thus the visit becomes an imaginary tour and the catalogue that comprises the *Narración* substitutes for a real one.

Was it appearance, artifice, or reality that was uppermost in Lastanosa's mind as he constructed this playful excursion into the nooks and crannies of his collection? Probably it was a combination of all three, for the imaginary tour was a kind of literary trope, a composite account that represented the many guided tours that Lastanosa gave to virtuosi and savants who made their way to his home. Surely the conceit was deliberate, and it gives us our first clue to understanding the meaning and importance of collections in courtly culture. The itinerary is a reflection of reality, just as the collection itself mirrors a cosmic reality.

The imaginary tour that Lastanosa guides will serve to introduce the subject of my paper. My purpose in this paper is to understand Lastanosa's strategy as a collector of 'secrets.' The irony that underlies that concept – collector of secrets – is, perhaps, obvious to the reader. Indeed, it is one of the central ironies of baroque culture, and I shall address that topic in the course of my investigation. At once a patron, courtier, and 'curioso,' Lastanosa was deeply immersed in the culture of 'virtuosity' and its fascination with the wonders and secrets of nature. Secrets occupied an important place in the collecting strategies of the courts, as they did in Lastanosa's collecting.

Appearance, artifice, and reality, the key words in my title, were central concepts that guided the collection and display of 'secrets' in baroque culture. Underlying the extensive collections of objects exhibited in the courtly culture of early modern Europe was a key premise about the discovery and understanding of the 'secrets of nature,' and this assumption provided the curiosi with a strategy for collecting secrets. The premise was that through artifice (i.e., by artificial means), one is led to discover a reality that lies hidden beneath appearances.[4] For nature has a two-fold aspect: it displays itself to our senses in the rich variety of the things of the world and, at the same time, conceals itself behind appearances in its most essential part. As Pierre Hadot writes, "The great secret of nature is Nature herself, that is, the invisible reason or force, of which the visible world is only the external manifestation."[5] How are nature's secrets made known? That, in many ways, was the great question of natural philosophy in the West, and it has been answered in a variety of different ways. In the Age of the Baroque, natural philosophers asserted that artifice alone would enable humans to probe and query nature and discover its secrets. Thus to Lastanosa and his contemporaries, artifice was not just a way of making and doing but a way of discovering and knowing.[6]

Let us then continue with our conceit and, in the manner of the baroque scholars that we celebrate in this book, accompany Lastanosa and the learned cleric on their imaginary tour.

Lastanosa doesn't want to bore his guest, nor disappoint him. Before proceeding, he warns the cleric that his vast collection (more than 2,000 volumes) may be fatiguing for a man who is not, like he, a 'curioso' and therefore interested in everything.[7] By identifying himself as a 'curioso,' Lastanosa meant something quite specific.[8] He intended to situate himself within a community of individuals who constituted themselves as a cosmopolitan elite that spanned national and confessional boundaries. Although time and wealth were one mark of the curioso's status, more distinctive was a painstaking attention trained on the novel, the rare, and the unusual events and objects of nature. The community of the curious, who also described themselves as the 'virtuosi,' was diverse, embracing physicians, clergymen, apothecaries, lawyers and merchants; but it was united in its preoccupation with the marvels of art and the secrets of nature. As Lorraine Daston and Katharine Park observe, "To count as one of the 'curious' was to combine a thirst to know with an appetite for marvels, which also came to be known as 'curiosities'."[9]

Collecting objects was deployed as a means of establishing one's credentials as a member of this new elite – an elite of wealth, to be sure, but also one of taste and refinement.[10] In courtly culture, collecting was related to the harnessing of wonder as a social and political asset.[11] Objects deemed suitable for collecting were deliberately selected for their anomalousness – thus setting them apart from the ordinary. The juxtaposition of objects with strikingly different themes and aspects emphasized the unusual, novel, and rare. Like a breathtaking display of wit, the exhibition of physical objects in collections and curiosity cabinets was designed to elicit wonder, both in regard to the objects and in regard to the collector of them.

"Aunque la curiosidad es virtud," Lastanosa modestly informs the cleric, "y prenda muy propia de los cavalleros que professan vivir retirados y empleados en algun honesto ejercicio, confieso que no la he afectado."[12] In fact, his curiosity was voracious and, apparently, almost completely indiscriminate. Yet his eclectic tastes were entirely typical of collectors of his day. Historians have offered a number of different explanations for the sudden explosion of the passion for collecting in early modern Europe, in particular among the nobility and aspiring nobility. Some historians stress the importance of the commercial revolution and, in particular, the growth of international trade, which made exotic naturalia and artificialia more readily available to collectors, whether through contacts with apothecaries or merchants.[13] Others have emphasized the political importance of collections as expressions of power and dominion over nature and the world, and have argued that representational prestige was enmeshed in a web of patronage and gift-giving, which had many diplomatic aspects.[14] Still others have linked collecting to the shifting fortunes of the nobility, arguing that the passion for collecting was a kind of pathology emanating from depression or *melancholia*, as the disease was known in the early modern period. Certainly intellectuals of the period across Europe thought that the disease was on the rise. Robert Burton, whose *Anatomy of Melancholy* (1621) became the defining treatise on the subject, detected an "epidemic" of melancholy in his time, particularly among the middling and upper classes.[15] In a recent article, Miguel López suggests that collecting was an aristocratic diversion aimed at alleviating melancholy, the malady of the baroque age, much in the manner described by Burton in his famous treatise.[16] It is an intriguing hypothesis, for it does seem that melancholy was a 'fashionable' disease and that certain writers cultivated aspects of the condition. Moreover, the tone of the *Narración* seems to portray a melancholic figure who isolates himself from a tense and troubled world and immerses himself in his collection of exotica. Clearly, more research needs to be done in order to properly assess the meaning of melancholy in the early modern period. What we can say is that the curiosi (or virtuosi, as they were also called) were, by and large, aristocrats and aspiring gentlemen (like Lastanosa) for whom collecting was an all-consuming passion and diversion.

Like most baroque collectors, Lastanosa made no firm distinction between the objects that he collected and the books that represented them. After all, texts were also collectible objects. Accompanying him on the tour of his house, passing from room to room, we encounter books, antiquities, medals, gems, scientific instruments, and strange and unusual 'rarities' all mingled in a free and playful assemblage of particulars. If anything gives this extraordinary collection unity – if there is any theme to be found in it – it is a preoccupation with rarity, novelty, the bizarre, and the wonderful. Disdaining the obvious and the merely useful, Lastanosa followed the convention of courtly collecting by riveting his curiosity on what was obscure, rare, and "marvelous." Lastanosa's entire collection is a testimony to a preoccupation with the marvelous, the only kind of objects that distinguished the virtuoso from the 'vulgar sort.' For this reason, 'secrets' – along with the books of secrets that represented and revealed them – occupy a central place in the collection.

It is not at all surprising that one of the first stops on Lastanosa's virtual tour was his extensive collection of the writings of the Jesuit polymath Athanasius Kircher. Lastanosa owned no less than a fifteen books by the erudite Roman priest. The books had, in fact, been sent by Kircher himself, and the case containing them is as good a place as any to begin our tour of Lastanosa's collection of 'secrets' and books of secrets.[17]

Here is not the place to try to decipher Kircher's dense and complex natural philosophy. However, I would like to call attention to one of Kircher's works, because it underscores a way of thinking about the 'secrets of nature' that was typical of baroque culture. The frontispiece to Kircher's *Magneticum Naturae Regnum* (Rome, 1667), one of his most famous works, proclaimed: "The world is bound with secret knots" (*Arcanus nodis ligantur mundus*).[18] For Kircher, magnetism was the chain that linked together all things in the universe. It explained phenomena that modern scientists would regard as quite unrelated – for instance, the sunflower's tendency to face the sun and the sexual life of animals. Magnetism was also the force that kept planets in their orbit, assuming the role that gravitation would later take in Newton's system. Friendship, love, hatred, chemical reactions, the qualities of medicinal plants and stones, the tides, and musical harmony were all governed by magnetism. Even the nature of God, whom Kircher called "the Central Magnet of the Universe," could be explained by magnetic force. Magnetism was to Kircher "the path to the treasure of the entire world" and a metaphor that stood for all the secret forces, sympathies, and antipathies that linked nature in an intricate web of correspondences.[19]

Most early modern cabinets of curiosities seem, on the surface, to have repudiated any concept of order. Lastanosa's is no exception. Carved gems, watches, antiques, mummies, and mechanical contrivances are displayed alongside fossils, shells, skeletons, giants' teeth, unicorns' horns, and exotic specimens from the New World, making up an encyclopedia of the bizarre and the marvelous. Full of strange and exotic *naturalia* and *mirabilia*, they seem intended only to delight spectators and to provoke wonder rather than to serve as museums for research. Yet, in fact, the bewildering variety of objects included in the collections, the deliberate blurring of the boundaries between art and nature, and the heightening of the experience of the anomalous, served to illustrate the baroque view that the multiformity of nature concealed a web of secret relations and forces: they are the 'secrets of nature' that natural philosophy was supposed to reveal.

The term, 'collector of secrets,' which I have used to describe Lastanosa, underscores a central irony about 'secrets' in the early modern era: On one hand, the possession of secrets distinguishes a virtuoso, setting him apart from the 'vulgar' who lack them. Yet if, on the other hand, the purpose of a courtly collection is to display one's virtuosity, then it must be open to a certain public. "[H]ay poco que ver en mi casa," Lastanosa assures the cleric.[20] Like almost all museums of the day, the collection was closed to all but a select few. Early modern collections were situated on the border between the private and the public, at once the property of an individual and requiring a public to validate the noteworthy status of the collector.

This was exactly the irony that confronted the sixteenth-century 'professors of secrets' who created the immensely popular genre of the printed books of secrets.[21] How do you both have a secret and, at the same time, reveal it – an act that makes the secret public? The dilemma is illustrated in the preface to one of the most famous books of secrets of the period, Alessio Piemontese's *Secreti*, which was first published in Italian in 1555 and appeared in over one hundred early modern editions and translations.[22] Lastanosa owned a 1624 Spanish translation of the work.[23] Alessio, the prototypical professor of secrets, hoarded up and jealously guarded his secrets, refusing to disclose them to anyone. "If everyone knew the secrets," he reasoned, "they would no longer be called secrets, but would be public and common."[24] Alessio paid the price for his vanity when his refusal to give away one of his secrets caused a man's death. Gripped by remorse, he resolved to publish them all. What Alessio and many others discovered was that, in the age of printing, publicity could be a far more effective means to gain fame than secrecy.

One solution to the moral dilemma posed by Alessio's preface was the catalogue. A peculiarly early modern literary genre, the catalogue was the result of the impulse not only to collect but to textualize collections. With the catalogue, one could claim ownership of secrets and even publicize them in tantalizing descriptions, and at the same time control access to them. The catalogue both represents a group of physical objects and is also, itself, an artifact – an object produced by a collector to demonstrate his propriety relationship to other objects.[25] By means of the catalogue (such as the one that Lastanosa had printed in 1647), the collector could both control access to his collection and 'display' them (virtually, of course) to a wider public. Hence, with the catalogue, the collection became an instrument of self-fashioning.[26]

Lastanosa had an impressive collection of books of secrets, both printed and manuscript. What did the "secrets of nature" mean to scholars and intellectuals of Lastanosa's age? The metaphor, ubiquitous in the early modern period, played upon several different senses of the concept of secrecy. One variety of secrecy was essentially *social*, involving the intentional suppression of information in order to protect knowledge from outsiders who might corrupt or abuse it, or to preserve the power that is inherent in private knowledge. An important *locus classicus* for this tradition in the medieval West was the pseudo-Aristotelian *Secretum secretorum*, which took the form of a letter from Aristotle to Alexander the Great.[27] The *Secretum* supposedly contained the wisdom that Aristotle reserved for his intimate disciples as opposed to the public doctrine that was studied in the schools.[28] The *Secretum* was thus a kind of manifesto of the doctrine of esotericism in medieval thought.[29]

Esotericism was also a characteristic of the ancient hermetic treatises, which attracted widespread interest in the Renaissance.[30] Lastanosa had several of the hermetic treatises in his library, and would also have been familiar with the tradition through his extensive collection of Kircher's *oeuvres*, where the ancient hermetic doctrine was well represented.[31] A fundamental ideological assumption behind this kind of esotericism is that knowledge is power, and that the concealment or control of knowledge is essential to the art of ruling. That, at least in part, explains Kircher's fascination with the art of secret communication, a theme that runs through all of his work. Whether magnetic, catoptric, musical, graphical, or numerical, all were tools to conceal secrets.

Another kind of secrecy was *epistemological*, implying that secrecy was a given in the order of nature, and that the "secrets of nature" are fundamentally unknowable: in other words, the conception of nature's secrets as *arcana naturae*. Central to it was the idea that nature is inherently arcane, and that the "veil" of nature's outer appearances conceals hidden qualities and virtues, the "secrets of nature" that lie within.[32] That concept was the core doctrine of the Hermetic occult sciences. In the Middle Ages, this was defined as the problem of occult qualities, and it had its classic expression in a work attributed to Albertus Magnus, the *Liber aggregationis*.[33] This work, very likely, was the "book of secrets" ascribed to Albert that is referred to in the *Narración*.[34] Ps-Albert's work described magical "experiments" that probed and manifested the occult qualities in nature, bringing to light nature's hidden secrets.

A third kind of secrecy implied by the "secrets of nature" metaphor was the one advanced in the Renaissance books of secrets, which are plentifully represented in Lastanosa's library. It might be called *epistemic* secrecy because it involved the recognition that the secrets of nature were not a given, not built into nature, but were an artifact of a way of looking at the world.[35] Craft secrets, for example, were secret only because few people knew them, while other secrets occurred purely by chance or, conversely, were the results of experiments aimed to tease out nature's arcana. One of the most famous works revealing such secrets was the immense encyclopedia of secrets by the German polymath Johann Jacob Wecker (1528-1586), *De secretis*, originally published in Latin in 1582 and reprinted many times thereafter. Lastanosa's edition of the work is not identified, but it was probably one of the Latin imprints.[36]

Wecker's *De secretis* was a hefty compilation of recipes, experiments, and observations extracted from classical, medieval, and contemporary authors. His list of 129 authorities was one of the most unusual assortments of authors ever assembled in the history of scholarship. In addition to familiar classical names such as Aristotle, Pliny, Galen, Ovid, and Virgil, it included several famous professors of secrets such as Alessio Piemontese, Leonardo Fioravanti, Gabriele Falloppio, and Giambattista Della Porta. The work took recipes from the German craft manuals called *Kunstbüchlein* and from the magical books of Hermes and Zoroaster. Modern authors such as Levinus Lemnius, Girolamo Cardano, and Paracelsus appeared alongside the secrets of Democritus, Orpheus, and Apollo's oracle. In Wecker's list of authorities, history seemed to demonstrate that the revelation of natural secrets was an ongoing, never-ending process. "The treasury of nature is inexhaustible," he exclaimed, "and undoubtedly much that lies hidden will with the efforts of sagacious men of succeeding times be uprooted."[37]

These works do not exhaust Lastanosa's sizeable collection of books of secrets. His library also included works by Fioravanti, Falloppio, and Cardano, as well as several manuscript books of secrets.[38] Among the latter is a manuscript identified as a "libro de la naturaleza de las piedras, traducido de griego, . . . y muchos otros secretos," probably a work related to the Hellenistic *Kuranides*.[39] Then there were works on craft and technical secrets, including Antoine Mizauld's *Secrets of Agriculture*, a text of obvious interest to a man who took such pride in his garden as did Lastanosa.[40] The garden, like many others of the day designed in the form of a vast labyrinth, was itself a metaphor, among other things, of the great winding search for the secrets of nature.

Nor were books the only way in which a curious collector might investigate the secrets of nature. "Acompañan estos libros muchos instrumentos en que se ven raras maravillas," explains Lastanosa.[41] One of the rooms in his house contained books and instruments on the science of catoptrics, a branch of mathematical magic. Catoptrics, the science of mirrors and reflected images, has a history going back to Euclid, Archimedes, and Hero of Alexandria.[42] In the Renaissance, it was one of the subjects that made up the immense field of natural magic, still in Lastanosa's day considered to be a legitimate scientific discipline. Natural magic endeavored to harness nature's occult forces and use them to produce 'marvelous' effects, like the powerful burning mirror that Archimedes supposedly used to set fire to the Roman navy that besieged

Syracuse. In scientific terms, such effects were 'marvelous' because they do not occur by nature, but happen only with the assistance of artifice.[43] Optical illusions and 'wondrous' images created by mirrors and lenses were among the most impressive achievements of natural magic. Art, science, and wonder intermingled, producing the typically strange and distorted images characteristic of the Baroque.

Alchemy, too, could assist the investigator in exploring the secrets of nature. Lastanosa professed to be dubious about the inflated claims of the alchemists. The visiting padre informs him that he has heard that the virtuoso was an adept. Lastanosa demurs, explaining, "aviendo tenido grandes ocasiones de saber dessa prodigiosa arte, creiendo que el fin de ella era enriquecer el hombre, siempre lo dexé despreciando esse modo de tesoros"[44] Yet, for a doubter, he was certainly fascinated by the subject.[45] As Bruce Moran points out in his contribution to this volume, Lastanosa owned a number of alchemical books, some allegorical and mystical, others practical, and he seems to have had a pronounced preference for the latter. His son, Vicente Antonio, in an éloge of his father rather quaintly titled *Habitación de las Musas*, proudly reported that Lastanosa translated the French chemist Beguin's *Elements of Chemistry* into Spanish.[46] The translation, evidently, was never published and the manuscript of the work (assuming that Vicente Antonio's account is accurate) did not survive the dispersal of Lastanosa's library and papers. Yet, whether or not the translation was completed, Vicente Antonio's testimony is compelling evidence of Lastanosa's alchemical interests. Beguin's famous textbook was essentially a compilation of chemical formulas.[47] Beguin's definition of alchemy – or, as he calls it, chemistry – as an art of making things aligned perfectly with Lastanosa's conception of using artifice to explore the secrets of nature. Chemistry, Moran writes, "offered a means of entry into nature's secrets because it described how to reduce things to their first principles."[48]

Lastanosa also owned a copy of one of the most famous early modern works on distillation: *Thesaurus Euonymi Philiatri de remediis secretis*, a work attributed to 'Euonymus Philiatrus' but in fact by the Swiss naturalist Conrad Gesner.[49] Gesner's great Latin work on distillation, originally published in 1552, was based on the idea that distillation enables the operator to separate the pure from the impure parts of a body and to extract the "secret arcana in order to make remedies known hitherto by very few persons, which empirics concealed as mysteries."[50] The seriousness of Lastanosa's interest in this subject can be gauged by the fact that, as we learn at the end of the *Narración*, some years prior to the learned cleric's visit, he had hired an alchemist to make quintessences. The mysterious

Neapolitan alchemist Natale Baronio came to Lastanosa with promises of making miraculous cures, obviously hoping to benefit from Lastanosa's well-known generosity as a patron. Baronio's visit seems to have quickened Lastanosa's enthusiasm for alchemy. In any case, the Italian produced something that Lastanosa was convinced was potable gold, a mere four drops of which could cure a man taken for dead.[51] Well, Natale Baronio was not the first Italian alchemist to fool a curious Spaniard. As Mar Rey Bueno has shown, Philip II's court played host to numerous foreign alchemists, many of whom were Italians.[52] It's not for nothing that the word for charlatan – *ciarlatano* – was originally an Italian word.

The books of secrets that do not show up in Lastanosa's library catalogue are almost as telling as those that appear. For example, he did not, evidently, have a very representative collection of the writings of the Neapolitan magus Giambattista Della Porta, whose voluminous and authoritative works on natural magic were generally regarded as essential texts of the tradition.[53] Nor, evidently, did Lastanos own a copy of the Flemish natural philosopher Levinus Lemnius's *Occulta naturae miracula*, another work that helped define the sixteenth-century approach to the science of occult qualities.[54] The absence of a copy of Lemnius's book might possibly be explained by the author's Protestant leanings – Lastanosa did not have any of Clusius's works either. In those days, it probably wasn't very prudent for a Spanish collector to own books by a Dutchman.

However, Della Porta's relative absence is a little more puzzling, considering that for a long time he was Europe's foremost authority on natural magic. A single reference to an unnamed Della Porta volume in the *Narración* probably refers to his famous book, *Magia naturalis*, but curiously, it is listed among a group of works devoted to "Tropelias [eutropelias] y Juegos de Manos," in other words, innocent amusements, sleight of hand, magic tricks, and so forth.[55] Does that mean that Lastanosa did not take Della Porta's brand of natural magic seriously? Certainly Della Porta's fascination with tricks and illusions has made it difficult for modern historians to take the *Magia naturalis* very seriously. The enthusiasm with which he described how to produce half-white, half-black figs, gigantic leeks, nuts without shells, sweet lemons, cucumbers shaped like dragons, to sire multicolored horses, counterfeit gemstones, or make an artificial egg as big as a man's head seems to result more from perverse curiosity than genuine scientific interest. But to Della Porta and his contemporaries, such amusements were not only prescribed by the rules of courtly discourse, they were also considered appropriate categories for the study of natural history. Nature's subtlety and 'playfulness' was thought to be an essential part of the architecture of the world.[56] Sometimes nature mimics itself, as in the seahorse or the mandrake, while at other times it contorts itself and exhibits monsters, giants, and dwarfs, or it demonstrates its cunning by creating artful imitations of human artifacts. So nature cloaks herself in masks, which it is the task of the naturalist to remove, thereby exposing nature's secrets. The illusions and sleights-of-hand that Della Porta included in his book were imitations of nature hiding herself and were part of the unmasking of nature that natural magic aimed to accomplish. Once exposed, nature might be imitated by the art of natural magic. But if nature 'plays' for those who understand her secrets, she also deceives those ignorant of causes, just as a juggler, magician, or craftsman seems to create marvels in the eyes of amazed onlookers. Revealing the mechanism underlying cunning, whether human or of nature, was an essential part of Della Porta's 'science of secrets.'

Of course, Lastanosa had plenty of natural magic for anyone's taste in his extensive collection of titles by Kircher and by his fellow Jesuit scholar Juan Eusebio Nieremberg – and perhaps therein lay the explanation for the apparent marginalization of Della Porta.[57] Perhaps it seemed to Lastanosa that the Jesuit natural philosophers, such as Kircher, Nieremberg, and the German Jesuit Gaspar Schott, propounded a deeper, more rigorous, and more orthodox form of natural magic than the playful and unstable version espoused by Della Porta. In the early seventeenth century, Della Porta's famous book would have been obligatory reading for anyone interested in the subject, and might have been shelved among other works on the occult sciences instead of among books about magic tricks and illusions. However, his reputation was clouded by his brushes with the Inquisition and his books were suspect: Della Porta had been brought before the Holy Office at least twice, and one of his books had been listed on the Index.[58] According to the eighteenth-century historian Giuseppe Valletta, who had the Inquisitorial documents before him, Della Porta was charged with "having written about the marvels and secrets of nature."[59]

Lastanosa's catalogue may thus be a sign that natural magic had moved on to become a more serious pursuit. Yet Lastanosa had an impressive collection of philosophical toys of his own, including examples of nature's sleights-of-hand in the form of figured stones that imitated a seemingly infinite variety of natural and artificial objects, as well as lodestones, amulets, mechanical marvels, magic mirrors, and all variety of scientific instruments.[60] Moreover, even in the dense Latin tomes of the Jesuits, the question of the boundaries of curiosity emerged as a central concern. Gaspar Schott wrote at the very beginning of his *Magia universalis* of

1657: "here the theatre where art and nature play is exposed to curiosity: but while they play for the learned, they deceive the ignorant," demonstrating that the divide between the elite and the vulgar still lay at the heart of natural philosophy.[61]

What, finally, does all of this passionate concern, in this quasi-courtly culture, with 'secrets' and experiments signify? To some extent it attests to a keen interest in applied science. Many of the recipes in the books of secrets recorded experimental attempts to improve artistic or technological processes. All of these would have been of great interest to a collector like Lastanosa, whose assemblage of paintings, ceramics, armor, medallions, and jewelry was renowned. However, beyond expressing the tastes and pastimes of wealthy connoisseurs, there was also a more serious purpose in the pursuit of secrets through collecting. First of all, the nearly endless parade of natural and artificial wonders that filled Lastanosa's house underscored the fact that the world was much vaster and more complex than Aristotle had ever imagined. So many secrets still awaited discovery. While the natural objects and marvels displayed what lay beyond traditional knowledge, antiquities, fossils, and figured stones illustrated what lay behind.

And so we return to appearance, artifice, and reality. In the rhythm of passing from one object to the next on our imaginary tour of Lastanosa's museum, we can detect the emergence of a new way of investigating nature that was characteristic of early modern science. Elsewhere I have called this mode of inquiry the 'epistemology of the hunt' and have argued that it contrasted sharply with the kind of philosophical knowledge handed down in the medieval Scholastic tradition.[62] The metaphor of science as a hunt, which occurs repeatedly in the scientific discourse of the sixteenth and seventeenth centuries, testifies to the emergence of a new conception of the aims and methods of science. Instead of viewing natural philosophy as a sort of hermeneutics – "natural philosophy without nature," as the historian John Murdoch aptly characterized late-medieval physics – intellectuals of the early modern period tended to think of science as a search for new and unknown facts, or of causes concealed beneath nature's exterior appearance.[63] Such methods, they argued, offered few insights into the changing, shifting world of becoming. Experiments alone would enable investigators to penetrate the vast regions of unknown nature. The hunt metaphor was used increasingly as natural philosophers attempted to elucidate and to vindicate experimentalism.

The idea of science as a hunt seems particularly appropriate to describe the museum-science of the early modern era.[64] Collecting focused the attention upon individual instances as opposed to the ascent to lofty principles. In amassing his immense collection of books and objects, Lastanosa did not aim principally to advance the cause of natural philosophy. His goals were more personal: to enhance his prestige and to project a self-image. Yet his museum and library, driven by a passionate dedication to collecting secrets, was a visual reminder that objects were clues that would lead to the discovery of nature's secrets.[65] Lastanosa was not, perhaps, a *novator*, and certainly no Baconian, but, like many other collectors of his day, in his own way he furthered the ideal of factuality that was of cardinal importance in the early stages of the Scientific Revolution.[66]

Notes

1. Ricardo del Arco y Garay, *La erudición aragonesa del siglo XVII en torno a Lastanosa*, (Madrid, Cuerpo facultativo de archiveros, bibliotecarios y arqueólogos, 1934), 252-75. The account is also published in *Narración de lo que le pasó a Don Vincencio Lastanosa a 15 de octubre del año 1662 con un religioso docto y grave*. Hispanic Society of America (HSA), Ms. B-2424, 52r-79vº.
2. On Lastanosa's circle, see Miguel López Pérez's contribution to this volume.
3. Ricardo del Arco y Garay, "Don Vincencio Juan de Lastanosa: Apuntes bio-bibliográficos," *Boletín de la Real Academia de la historia*, 56 (1910): 420-511.
4. On Lastanosa's fascination with the secrets of nature and art, see María Portuondo's contribution to this volume.
5. Pierre Hadot, *The Veil of Isis: An Essay on the History of the Idea of Nature*, (Cambridge, Harvard University Press, 2006), 34.
6. Antonio Pérez-Ramos, *Francis Bacon's Idea of Science and the Maker's Knowledge Tradition*, (Oxford, Oxford University Press, 1988). Pérez-Ramos calls this way of discovery 'maker's knowledge.' In addition, see Pamela Smith, *The Body of the Artisan: Art and Experience in the Scientific Revolution*, (Chicago, University of Chicago Press, 2004).
7. *Narración*, 52v.
8. Katie Whitaker, "The Culture of Curiosity," in: Nicholas Jardine, James A. Secord y Emma C. Spary, eds., *Cultures of Natural History*, (Cambridge, Cambridge University Press, 2000), 75-90; Krzysztof Pomian, *Collectors and Curiosities: Paris and Venice 1500-1800*, (Cambridge, Blackwell, 1990).
9. Lorraine Daston and Katharine Park, *Wonders and the Order of Nature 1150-1750*, (New York, Zone Books, 1998), 218.
10. Richard A. Goldthwaite, *Wealth and the Demand for Art in Italy, 1300-1600*, (Baltimore, Johns Hopkins University Press, 1993), 249.
11. Paula Findlen, "Courting Nature," in: Jardine, Secord & Spary, *Cultures of Natural History*, 57-74.
12. *Narración*, 52r.
13. Harold J. Cook, *Matters of Exchange: Commerce, Medicine, and Science in the Dutch Golden Age*, (New Haven, Yale University Press, 2007). On pharmacies as

sites for collecting, see Paula Findlen, *Possessing Nature: Museums, Collecting, and Scientific Culture in Early Modern Italy*, (Berkeley, University of California Press, 1994), 241-87.

14. Thomas DaCosta Kaufmann, "From Mastery of the World to Mastery of Nature: The Kunstkämmer, Politics, and Science, in: *The Mastery of Nature: Aspects of Art, Science, and Humanism in the Renaissance*, (Princeton, Princeton University Press, 1993), 174-94.

15. Angus Rowland, "The Problem of Early Modern Melancholy," *Past and Present*, 191 (2006): 77-120 (p. 77).

16. Miguel López Pérez, "Anatomía del virtuoso: coleccionismo y melancholia en la figura de Vincencio Juan de Lastanosa," *Argensola*, 115 (2005): 143-66. In addition, see Lorraine Daston, "Curiosity and Early Modern Science," *Word and Image*,11 (1995): 391-404.

17. Lastanosa had exchanged letters with Kircher since the 1650s; as a result, in 1660, Kircher sent him 15 of his books: Carlos Garcés Manau, "Diez cartas de Vincencio Juan de Lastanosa y Diego Vincencio Vidania a Athanasius Kircher, conservadas en la Universidad Pontificia Gregoriana de Roma," *Argensola*, 115 (2005): 187-99 (pp. 189-90).

18. Although this particular book does not appear in Lastanosa's library catalogue, he did own a copy of Kircher's *Magnes, sive de arte magnetica* (Rome, 1652), which contains similar ideas. On Kircher's *Magnes* and the centralisty of magnetism in his natural philosophy, see Paula Findlen, "The Last Man Who Knew Everything…or Did He? Athanasius Kircher, S.J. (1602-80)", in: Paula FINDLEN, ed., *Athanasius Kircher: The Last Man Who Knew Everything*, (New York, Routledge, 2004), 1-48. The frontispiece to Kircher's *Magneticum Naturae Regnum* is pictured in Ingrid D. Rowland, *The Ecstatic Journey: Athanasius Kircher in Baroque Rome*, (Chicago, University of Chicago Press, 2000), 37.

19. Quoted in Findlen, *Possessing Nature*, 85.

20. *Narración*, 52r.

21. William Eamon, *Science and the Secrets of Nature: Books of Secrets in Medieval and Early Modern Culture*, (Princeton, Princeton University Press, 1994).

22. On this work, see Eamon, *Science and the Secrets of Nature*, 139-47.

23. *Catálogo de los libros. D. Vincencio Ioan Lastanosa. Por orden de alfabeto.* Stokholm Royal Library, *Sparvenfeldt* Catalogue (K. B. Sp. 10-U 379), 1v. According to the catalogue, Lastanosa's was the Madrid, 1624 edition. This edition was evidently unknown to Ferguson and is not listed in his bibliography of books of secrets or his article on Alessio: John K. Ferguson, *Bibliographical Notes on Histories of Inventions and Books of Secrets*, 2 vols., (London, Holland Press, 1959); idem "The Secrets of Alexis," *Proceedings of the Royal Society of Medicine, Section on the History of Medicine*, 24 (1931): 225-46.

24. PIEMONTESE, Alessio Piemontese, *Secreti*, (Venice, 1558), preface. The preface containing the story about Alessio's decision to publish the secrets is not contained in the Spanish translation, but does appear in most other translations, including the Latin, and was widely known throughout Europe. Given his extensive contacts, Lastanosa was undoubtedly familiar with the legend.

25. Marjorie Swann, *Curiosities and Texts: The Culture of Collecting in Early Modern England*, (Philadelphia, University of Pennsylvania Press, 2001), 7. In addition, see Lorraine Daston, "The Factual Sensibility," *Isis*, 79 (1988): 452-67 (pp. 459-60).
26. The historian Richard Lightbown reported that in 1647, Lastanosa had printed 2,000 copies of a verse description of his collection by Juan Francisco Andrés de Uztarroz to satisfy requests by those curious about his museum. However, I have not been able to discover any copies of the publication. Richard Lightbown, "Some Notes on Spanish Baroque Collectors," in: Oliver Impey and Arthur MacGregor, eds., *The Origins of Museums: The Cabinet of Curiosities in Sixteenth- and Seventeenth-Century Europe*, (Oxford, Clarendon Press, 1985), 136-46 (p. 144).
27. On the *Secretum*, see Steven J. Williams, *The Secret of Secrets. The Scholarly Career of a Pseudo-Aristotelian Text in the Latin Middle Ages*, (Ann Arbor, University of Michigan Press, 2003). In addition, see William Eamon, "The Secrets of Nature and the Moral Economy of Early Modern Science," In: *Il Segreto / The Secret*, A. Paravacini Bagliani (ed.), *Micrologus*, XIV (2006), Florence, SISMEL, 215-35.
28. The work's Arabic translator, Yahya ibn al-Batriq, claims he found the manuscript in a temple of Asclepius after wandering through Persia in search of secret wisdomWilliams, *The Secret of Secrets*, 9.
29. On the esoteric tradition in medieval thought and the *libri secretorum*, see Eamon, *Science and the Secrets of Nature*, chap. 2.
30. There is a large literature on Renaissance hermeticism; see, in particular, Brian Copenhaver, "Natural Magic, Hermetism and Occultism in Early Modern Science," in: D.C. Lindberg y R.S. Westman, eds., *Reappraisals of the Scientific Revolution*, (Cambridge, Cambridge University Press, 1990), 261-301.
31. Noel Malcom, "Private and Public Knowledge", in: Findlen, *Athanasius Kircher*, 297-308. The work identified in Lastanosa's library catalog as a "libro de la naturaleza de las piedras, traducido de griego" was probably a treatise related to the Hellenistic *Kuranides*, a hermetic treatise on the virtues of natural objects (*Catálogo*, 117v). On the hermetic treatises, see Brian Copenhaver, *Hermetica*, (Cambridge, Cambridge University Press, 1992) (on *Kuranides*, pp. xxxiv-xxxv); and Garth Fowden, *The Egyptian Hermes: A Historical Approach to the Late Pagan Mind*, (Princeton, Princeton University Press, 1986).
32. The concept is treated in depth in Hadot, *The veil of Isis*.
33. On this work, see Eamon, *Science and the Secrets of Nature*, 71-74.
34. *Narración*, 59v.
35. I use the term "epistemic" in the sense elaborated by Michel Foucault, *The Order of Things*, (New York, Vintage Books, 1973).
36. *Narración*, 59v.
37. For additional information on Wecker and his *De secretis*, see Eamon, *Science and the Secrets of Nature*, 276-78.
38. *Narración*, 59v.
39. "Un libro de la naturaleza de las piedras, traducido de griego por un obispo de Rodas, y muchos otros secretos. En 4°." *Catálogo*, 105r.
40. *Catálogo*, 117v.

41. *Narración*, 55v.
42. For the scientific background to catoptrics in the Renaissance, see Sven Dupré, "The Dioptrics of Refractive Dials in the Sixteenth Century," *Nuncius*,18 (2203): 39-67.
43. Giambattista Della Porta, the principal Renaissance authority on natural magic, treated catoptrics in his famous work on natural magic: Giambattista Della Porta, *Magiae naturalis libri viginti*, (Naples, 1589), Book XVII, 259-86 ("De catoptrica imaginibus"). His ideas are discussed in Vasco Ronchi, *The Nature of Light*, (London, Heinemann, 1970), 78-87.
44. *Narración*, 78v-79r.
45. On Lastanosa's alchemical interests, see Miguel López-Pérez, *Asclepio renovado: Alquímia y medicina en la Expaña moderna (1500-1700)*, (Madrid, Ediciones Corona Borealis, 2003), 268-75.
46. *Habitación*, 2v.
47. On this work, see Bruce Moran, *Distilling Knowledge: Alchemy, Chemistry, and the Scientific Revolution*, (Cambridge, Harvard University Press, 2005), 112-14.
48. Ibid., 117.
49. Identified in the *Narración* as 'Filatro,' meaning Euonymus Philiatrus, Gesner's pseudonym.
50. On Gesner, see Lynn Thorndike, *History of Magic and Experimental Science*, vol. 5, (New York, Columbia University Press, 1941), 621.
51. *Narración*, 77v.
52. Mar Rey Bueno, *Los señores del fuego: Destiladores y espagíricos en la corte de los Austrias*, (Madrid, Ediciones Corona Borealis, 2002).
53. On Della Porta, see Eamon, *Science and the Secrets of Nature*, 194-233.
54. On this work, see Eamon, *Science and the Secrets of Nature*, 274-75.
55. There has been a considerable amount of attention to the word eutropelia in Cervantes; see, in particular Bruce W. Wardropper, "La eutrapelia en las *Novelas ejemplares* of Cervantes," in: *Actas del Séptimo Congreso Internacional de Hispanistas*, (Rome, Bulzoni, 1982), 153-69; Cory A. Reed, "Ludic Revelations in the Enchanted Head Episode in Don Quijote (II, 62)," *Cervantes: Bulletin of the Cervantes Society of America*, 24 (2004): 189-216; Joseph R. Jones, "Cervantes y la virtud de la eutrapelia: la moralidad de la literatura de esparcimiento," *Anales Cervantinos*, 23 (1985): 19-30.
56. Paula Findlen, "Jokes of Nature and Jokes of Knowledge: The Playfulness of Scientific Discourse in Early Modern Europe," *Renaissance Quarterly*, 43 (1990): 292-331.
57. Lastanosa also did not have a copy of Girolamo Cardano's famous *De subtilitate*, another key sixteenth-century text on occult qualities (although he had a copy of Cardano's *De sapiente et consolatione*); but he did have several works by Nieremberg, including *Curiosa filosofía y tesoro de marauillas de la naturaleza examinadas en varias questiones naturalis.*
58. On Della Porta's case before the Inquisition, see Giovanni Aquilecchia, "Appunti su G. B. Della Porta e l'inquisizione," *Studi secenteschi* 9 (1968): 3-31;

Pasquale López, *Inquisizione stampa e censura nel regno di Napoli tra '500 e '600*, (Naples, Edizioni del Delfino, 1974), 153-59.
59. "per avere scritto intorno alle maraviglie et i segreti della natura," quoted in Aquilecchia, "Appunti", 6.
60. *Narración*, 72r-v.
61. Quoted in Rob Iliffe, "Lying Wonders and Juggling Tricks: Religion, Nature and Imposture in Early Modern England," en: James E. Force and David S. Katz, eds., *Everything Connects: In Conference with R.H. Popkin. Essays in His Honor*, (Leiden, Brill, 1999), 185-209 (p. 196).
62. Eamon, *Science and the Secrets of Nature*, 269-300; in addition, see William Eamon, "Science as a Hunt," *Physis* 31(1994): 393-432.
63. John Murdoch, "The Analytical Character of Medieval Learning: Natural Philosophy Without Nature," in *Approaches to Nature in the Middle Ages*, L. D. Roberts (ed.), Medieval and Renaissance Texts and Studies, 16 (1992), (Binghampton, N.Y., Center for Medieval and Renaissance Studies), 171-213. In addition, see Paolo Rossi, "The Aristotelians and the 'Moderns': Hypothesis and Nature," *Annali dell'Istituto e Museo di Storia della Scienza di Firenze*, (1982): 73-28.
64. I use the term 'museum science' in the sense coined by Ken Arnold, *Cabinets For The Curious: Looking Back At Early English Museums,* (London, Ashgate, 2006).
65. On Lastanosa's "afición a secretos," see José María López Piñero, *Ciencia y técnica en la sociedad española de los siglos XVI y XVII*, (Barcelona, Labor, 1979), 410-11.
66. Daston, "The Factual Sensibility," 464.

CHAPTER NINE

EXTRACTING THE VIRTUES OF NATURE: SPAGYRIC REMEDIES AND CHEMICAL METAPHORS IN THE LIBRARY OF DON VINCENCIO JUAN DE LASTANOSA (1607-1681)

BRUCE T MORAN

Records from the library of Vincencio Juan de Lastanosa document the existence of a curious assembly of alchemical books and books related to the making of chemical medicines. Some of the best known authors blending alchemy and medicine in the early modern era, including Oswald Croll (1560-1609), Jean Beguin (1550-1620), and Philip Ulstadius (fl. 1525),[1] were part of that collection. Their works have been frequently discussed before, and rather than concentrating upon them yet again, the following study focuses upon the content of a few other books in the Lastanosa library that may give us a further clue as to why a local Spanish gentleman would wish to possess them as part of a larger medley of instruments, globes, maps, antiquities, and general curiosities. As we will see, some of these books described practical procedures by means of which certain healing powers could be drawn forth from plants, animals, minerals, and metals. However, in addition to such practical writings, other books in the library had a more erudite and theoretical purpose. In their pages something else was hidden and required skillful extraction. Rather than illustrating physical powers encased in natural objects, these texts represented intellectual insights veiled in metaphor and enigmatic narrative. Both sorts of text promised secrets, revealed either through understanding the procedures of chemical preparation or by means of properly interpreting symbol and allegory.

The history of alchemy in Renaissance and Early Modern Spain has recently received much deserved attention[2] and the Lastanosa library stands as a good reminder of the degree to which alchemical interests, combined with other cultural pursuits related to collecting and literary patronage, extended even to the relatively modest domain of a wealthy local patrician in upper Aragon. That Lastanosa was a vigorous collector of texts, *naturalia*, and *artificialia* is well known. One might quarrel about the extent of his knowledge concerning the things he assembled. However collecting without knowing much about what one collected would certainly have been considered the work of a fool -- and Lastanosa, whatever his difficulties in life, was no fool. Aside from the mere presence of specific books in the library, therefore, it is a good guess that he knew something about what was in them. In this regard, among the library's more allegorical writings, Vincencio Juan may have paid particular attention to the popular *Musaeum Hermeticum* with its ample supply of poetic and metaphorical references to nature and alchemy.[3] There he had access to the emblematic poem describing the powers of the Philosophers' Stone called *The Book of Lambspring*. Another text in the same collection called the *Book of Alze*, described allegorically the importance of Mercury in the process of producing a Grand Elixir. Still another tract, Jean de Meung's *The Remonstrance of Nature* described poetically the powers and virtues of potable gold. Each of these texts was included in the *Musaeum Hermeticum* and is referred to individually in existing records from the house of Lastanosa. Each contained secrets buried not in nature but in metaphor. If, then, the texts catalogued from his library are any indication, Vincencio Juan de Lastanosa may have combined intellectual interests of two sorts when collecting practical procedures related to making medicinal remedies and when seeking a knowledge of nature's powers through textual exposition. We might call these different sorts of interest, following a structure set out long ago by Jürgen Habermas,[4] curiosity based in technical control on the one hand, and curiosity linked to interpretative understanding on the other. Virtuous extractions, in other words, could be of both the practical and poetic sort. In giving attention to both, my focus will first be upon an important but little known pharmaceutical book in the Lastanosa library written by a German physician and nobleman named Adrian von Mynsicht (1603-1638). Thereafter the discussion shifts to other texts in the collection that may have been appealing more for what intellectual secrets they contained than for their descriptions of unusual or exotic medicinal preparations. In this regard the focus turns to texts written by two well known alchemical figures in the late sixteenth and early seventeenth centuries, Giovanni Battista Nasari and Pierre Jean Fabre (1588-1658).

Adrian von Mynsicht was a German alchemist, poet, count palatine, and personal physician to the duke of Mecklenburg, Adolph Friedrich. His real name was probably Adrian Seumenicht, which he re-stylized later, following his reception of a noble title, as Hadrianus a Munsicht, or Mynsicht. Regardless of the name, Mynsicht wrote one of the most popular books concerning the making of chemical medicines and the extraction of medicinal virtues in the early modern European world. The book, possessed by thousands of people and in print for almost a hundred years, was called in Latin *Thesaurus et Armamentarium Medico-Chymicum*, and first appeared sometime around 1630.[5] It was a huge success, with at least twenty-eight separate printings in thirteen different cities. Separate editions appeared at Rouen, Hanau, Hamburg, Tübingen, Offenbach, Amsterdam, and London. Three editions appeared in Lubeck, six in Lyon, three at Venice, three at Geneva, three at Stuttgart, and three more at Naples. Ten separate editions appeared in the early eighteenth century, the last in 1726. By then, both an English and several editions of a German translation had been available for many years. The library of Lastanosa possessed a relatively early copy, one printed at Lyon in 1640.

In some ways the *Thesaurus* is a typical work promoting new medicines as secrets made public for the good of human kind. The recipes and claims to specific medicinal virtues contained in the book, Mynsicht proclaimed, had been distilled from three sources. The first was his own personal experience; the second was a pharmaceutical understanding acquired through the "Light of Nature" (that is, by means of divine inspiration in recognizing which parts of nature related best to maintaining the health of specific parts of the human body); and, finally, from information divulged to him as a result of the good will of other spagyric artists.[6]

From those sources Mynsicht brought together recipes for medicaments made from a wide array of plant, animal, and mineral ingredients. Some were local and easily accessible. Others were foreign and hard to obtain. Some recipes required great wealth to produce. Others could not be had at any price and were more rumors than realities. One powerful remedy, the Bezoar of Jupiter, [made from regulus of antimony, English tin, sublimate mercury, and spirit of nitre] he described having first made himself in 1610 finding it "strong and powerful in operation." Among its virtues, the medicine worked wonderfully "in the plague, fevers. . . and all epidemical diseases," and was a "remedy against poison of any kind that is drunk."[7] Another medicament, Mineral Unicorn, he described as a universal medicine "able to cure all diseases" and, when mixed with the "pure gold of the philosophers," capable of perfecting all metals.[8] A further extraction, Solar Unicorn, by whose virtues life itself could be extended and the whole human body revived "though half dead" was such a mystery that Mynsicht was forced to admit that "it is not improper to call it a most powerful and inaccessible remedy."[9]

These were medicines meant for "princes and great persons." With certain alterations of ingredients, however, similar virtues could be extracted from nature at much reduced cost. His description of "royal cathartic," for instance, required the use of a "confection of alkermes," a secret composition invented specifically for "dainty constitutions." But ordinary and common folk, he advised, could substitute a conserve of vitriolated violets and get the same effect.[10] The substitution of parts was a common, but often secretive, practice among early modern pharmacists. Nevertheless, rare and closer-to-hand components were sometimes mixed into the same remedy. In this way hard to find mineral ingredients got mixed together with herbal substances that also contained active powers. Skilled apothecaries would have noticed that in Mynsicht's cathartics mineral purgatives like spirit of vitriol, prepared nitre, and vitriolated tartar were combined with parts of plants like the roots of scammony or the leaves of senna – then as now, among the strongest of vegetable cathartics in nature or the garden.

Miguel López Pérez has shown us that procedures derived from alchemical texts were becoming increasingly part of commercial practice and popular culture in seventeenth century Spain.[11] Money was certainly to be made from spagyrically prepared chemical medicines and such remedies were accessible usually only to those with lots of money to spend. The essences, tinctures, extracts, and magisteries described in Mynsicht's text and in many other pharmaceutical manuals were costly and usually far removed from the means of the average person. When the Leipzig physician, Martin Pansa, described his remedies in a book called *Pharmacotheca Publica et Privata* (1622-23) he reserved chemical medicaments for the apothecaries of "princes, nobles, and other persons of wealth."[12] The medicaments reserved for the socially and economically privileged promised to cure almost anything, as long as God did not have other plans. Pansa prepared a separate section of his book focused upon Galenical remedies. These were of lesser virtue but within the means of the earnest *Hausvater* and proper *Burger*. Finally in the last part of his text, Pansa described a *thesaurus pauperum* composed of remedies for "the poor, common man." These were herbal remedies made from ingredients of little or no cost. They possessed fewer virtues but were still good in several sorts of illness.

Pansa simply described a social-medical fact. The aristocracy and the rich could afford the most powerful medicines – if ingredients were at hand, and if someone could be found who knew how to prepare them. The house of Lastanosa was no exception in this regard. Diego Bercebal records that *laudanum opiatum* was prepared there by Don Nadal Varonio Veneciano, who also provided Lastanosa with the solvent of gold. Bercebal also mentions that attention was given there to the "Boli" of Adrian made from an extraction of opium (a bolus is a kind of soft paste that one lets melt in the mouth).[13] The reference is certainly to "Adrian" Mynsicht who described twelve separate preparations of boles made from *laudanum opiatum* flavored with a variety of mostly herbal ingredients.[14] It seems reasonable to assume that one or several of these remedies were attempted also at the house of Lastanosa. However, some recipes were certainly far from possible—there or anywhere else. Bercebal records, for instance, that one could look to Mynsicht for medicines that made use of the healing powers of pearls.[15] Indeed, oriental pearls were required by Mynsicht for his *confectio opiata.* But the confection also required, besides spagyrically prepared opium and red coral, foreign mummy, as well as unicorn horn.[16] Foreign mummy (usually the powdered remains of desiccated bodies unearthed in Egypt or elsewhere in the middle east) found its way into other recipes in Mynsicht's book including a "compound powder of mummy" recommended as a means of dissolving coagulated blood, and therefore as a way to prevent abscesses resulting from "falls from on high."[17] An epileptic powder required "the skull of a man strangled,"[18] while an equally remote medicament, "Aurea Hadrianea," (possibly Adrian's own concoction) was to be made from magistery of pearls and red corals, hermetically calcined hart's horn, elk's hoof, stag's heart, vipers, sealed earth, unicorn horn, and other indistinct preparations like Mithridate and Venetian Treacle. Its virtues were as exorbitant as its ingredients. Mynsicht declared that "whosoever in time of plague shall use this medicine, keep it in their bodies, and sweat with it, do by the Providence of God escape, without having the distemper."[19]

Beyond the grand promises that Mynsicht made for the virtues contained in his remedies, there were other reasons for readers to take an interest in what he wrote about. His recipes not only mixed herbal and mineral ingredients, but sometimes interlaced local and exotic materials as well. Substances from the new world gained special attention. The gum of Peru, prepared with spirit of vitriol, produced a red extract excellent for treating dropsy and gout.[20] Balsam of Peru, mixed with ambergris, oil of rhodium and calcined ivory made a *suffimentum angelicum* that produced a celestial odor capable of refreshing the brain and preventing "the contagious infection of a malignant air." It was also used as a perfume by kings, princes, and great men.[21] The tincture of Guaiac, extracted by means of spirit of wine (alcohol) produced virtues for drying catarrhs and taking away "the stink and rottenness of the mouth." A tincture of Sassafras made by infusing sassafras chips in water released virtues for countering arthritic pains. It also expelled the stone and prepared the womb for conception.[22]

The virtues of these ingredients, gathered from places far away, could not, however, come close to matching the virtues of another exotic material, one found both in New Spain and in the orient. This was bezoar, or more specifically, bezoar stone, a universal antidote thought to grow inside the bodies of certain ruminating animals, especially goats and deer, in India and, as reported by more contemporary explorers, also in the mountains of Peru. Mynsicht reported the preparation of several recipes that had either bezoar as the prime ingredient or in which the stone was added to other materials. The two most important were potable bezoar and magistery of the bezoar stone. Potable bezoar he regarded as an *arcanum* that was produced my mixing powdered bezoar stone with a powerful (and undisclosed) dissolving *menstruum*. He had himself administered the medicine, he notes, "with wonderful success among the infected, in acute diseases including the plague." The magistery of bezoar was prepared in the same way, except that after mixing the powdered bezoar stone with the *menstruum*, Mynsicht slowly added, drop by drop, juice of citron and distilled vinegar. The procedure produced "soft and subtle particles" that precipitated at the bottom of the vessel. These, when dried, resulted in a true magistery of bezoar whose virtues especially worked "against all poisonous humors or vapors that infect the heart."[23]

There were lots of stories about the effects of the bezoar stone. Mattioli praised it, saying that when one bound the stone to the left breast, it would drive away all poison from the body. Fallopius recorded the use of the stone as a medicament in combating plague at Ferrara. Antoine Mizauld described how some stones were thought to result from thickened tears under the eyelids of certain deer who plunged their heads in water after eating snakes or other venomous things (driven to do so by the growth of worms in their intestines and stomachs). At Goa, Garcia de Orta found a red deer in whose stomach grew such a stone, like an onion encased in skin. Nicolas Monardes wrote that some stones were round, and some elongated and some looked to him like the eggs of wild doves. Another Portugese, Pedro de Osma, spent five days in 1568 in the mountains of Peru hunting animals that might contain the stone. An Indian boy, ten or twelve years old, finally showed him where the stone could be found in certain animals and paid with his life for revealing such knowledge, because, de Osma reported, the Indians of that region also valued the stone highly and did not want its presence there known.[24]

The bezoar stone must have been popular in Aragon because it was popular everywhere else. The stone fetched a high price and continued to be in demand all over Europe well into the eighteenth century. In 1715, when the English chemist and physician Frederick Slare (1647?-1727) reported to the Royal Society about the effects of the stone, he noted that a friend and druggist in London sold around five hundred ounces of the stone every year. Weighing the stones, he determined that on average twelve stones made up one ounce and estimated therefore that more than five thousand animals needed to be slain to satisfy the annual demand for stones by just one pharmacist in London.[25] Slare wanted to make the point that regardless of what one believed about the virtues of such stones, fake stones were abundant. However, people had known that for a long time and it had made no difference at all to the stones reputation or desirability. In 1589 the German physician, Johann Wittich (1537-1598), noted that one could find lots of such stones for sale at Lisbon. However merchants there were loath to allow buyers to test them. What one needed to do, if given the chance, he advised, was to use a needle to pull a thread through a poisonous plant and then use the same needle to puncture the foot of a dog or some other animal. The thread attached to the needle could then be left in the wound. If, when the dog seemed to approach death, a bit of the stone was shaved off, mixed with water, and then the water poured onto the wound, the stone could be thought virtuous if the dog regained its strength. If the dog died, the stone was false.[26]

And then there was yet another reason why so many people, including I am sure Vincencio Juan de Lastanosa, liked to collect Mynsicht's book. In almost every one of the many editions of the text there was attached an appendix called the *Testament of Adrian Concerning the Gold of the Philosophers' Stone*. The *Testament*, however, followed from, and indeed helped to interpret, an earlier poetical text called the *Aureum Seculum Redivivum*, or *The Golden Age Restored*. That story had appeared in the *Musaeum Hermeticum* (1625) and was also written by Adrian von Mynsicht under the cover name Henricus Madathanus.[27]

In the story, a narrator begins by meditating upon the *locus classicus* for the description of potable gold, namely, the story of how Moses ground the idolatrous golden calf into powder, cast it upon the waters, and gave the golden water to the children of Israel to drink. When the storyteller falls asleep Solomon comes to him in a dream and displays to him a nude woman (the symbol of nature) from a wound in whose breast there flows blood and water and whose garments lie unsightly and poisonous at her feet. Although he is told that she has been used everywhere among the many people of the streets, it is she that the storyteller wishes to be his beloved. Give me this woman, the narrator says, although her garments are filthy and torn, "I will clean them and love her with all my heart." When he awakes the woman is gone, but the filthy robes remain. The garments, however, repulse him so much that he can't stand to touch them. For five years they remain as he found them. Then, in a second dream, an old woman (the mother of his beloved) appears, accusing him of the death of her daughter and scolding him for doing nothing with the remaining garments. But he had known nothing of her death, the dreamer insists, and had never learned the use of the garments left behind. The old woman then explains that her daughter (Nature) has not really died, but has been transformed, and reveals further that she, motivated by a special love and affection, has left to him a grey marbled casket made fetid by the stench and dirt of the garments her daughter once wore, and which the narrator is instructed to clean. Doing so, Nature's beloved finally understands how to produce the virtues by which Nature was herself made lovely and strong. The same knowledge informs him of how the heart, and all members of the body, may be renewed by the same powers, how age can be driven away, how weakness can be destroyed, and how the entire world may be refreshed.

If stories like this one were appealing at the house of Lastanosa, they were engaging not as figurative descriptions of medicinal virtues submerged in material dross, but as representations of secret knowledge cloaked in enigma and fable. In just this way Vincenio Juan's copy of Giovanni Battista Nazari's well known *Three dreams of metallic transmutation* (1599)[28] contained an abundance of hidden wisdom concealed in the images of maidens, nymphs, dancing monkeys, multi-headed fantasies, and a monstrous hen that was also a monstrous rooster. In one instance, nature's active virtues of generation, multiplication, and perfection were depicted by Nazari as three fathers emerging from an original chaos. These produced in natural things powers capable of destroying any illness. In another image, body and soul were joined together to represent the entire operation of reconciling opposites and perfecting those things that were imperfect. "I kill death that killed me," the figure says. "I revive the bodies that live in death." "I am mother, father, and son. I am invisible when flying and untouchable when I run in the air. But if you would nourish me with fire, my body would change into another form, purified of mortal poisons. Then you will see body, soul, and spirit joined and you will do great things in the world. . ."[29]

Extracting divine virtues, including moral virtues, from the study of nature characterizes much of the writing also of Pierre Jean Fabre, the best represented alchemical author of the Lastanosa collection. Vincencio Juan's library boasted of every major title in the Fabre *corpus*, ten volumes in all. These books combined natural philosophy, hermetic cosmology, alchemy, and medicine. For Fabre, alchemy amounted to true philosophy since only alchemy was capable of inquiring into the causes and principles of created things, and, by means of its own fire and its own procedures, was able to reduce all things to the matter and principles from which they were first composed. Alchemy thus excelled all other forms of natural knowledge as it applied itself to know the physical world from the inside out, materially and formally, rather than simplistically on the basis of outside appearance.[30] In all his texts, one message especially stood out – that nature was not the craftsman of itself and that by studying nature through alchemy one could extract from the things of the world true philosophy and a knowledge of God.[31]

In books with titles like *Propugnaculum alchymiae*, *Palladium spagyricum*, *Alchymista christianus*, *Myrothecium spagyricum sive pharmacopoea chymica*, and *Hercules Piochymicus* (all found in the Lastanosa library) Fabre described how, through the study of alchemy, one could penetrate the outward corporeal shell of nature so as to grasp the spirit *in* nature (the spirit of the world that he called Hercules)[32] – an activating force derived from the stars and transmitted into bodies by light.

> "Light is the proper and true Form of the heavens. Light is also the substance of the heavens . . . from which the heavens has its being, as well as all its virtues and powers. . . So all the virtues and powers that pour forth from the heavens depend upon and pour forth from light alone, as from its simple beginning, since without the light of the heavens there would be no heavens at all. Take away light, then you have also to remove the heavens and everything else [because]. . . light itself is the Form of every body [and] . . . as the Form is the source of all action, so . . . if light or the innate heat is put out. . . then all activity would cease and death would follow."[33]

According to Fabre, the study of light through its spiritual representations was the glory, virtue, and ornament among all chymists. It offered a way of interpreting natural mysteries, and, when linked to alchemy and theology, taught as well moral philosophy.[34]

Understanding which part of the heavens influenced, by its rays, specific parts of the terrestrial world was the true purpose of natural inquiry and the true activity of the alchemist. "Although lying hidden invisible and in a spiritual manner in all bodies, the spirit is made visible and ready to hand by an experienced chymist, so that that which lies spiritually and invisible in bodies is [made] visible and corporeal."[35] Extracting celestial virtues from nature required the alchemist, with the help of Vulcan, to extract an essential salt (the body of nature in which all powers lay hidden) from minerals and metals. "When," Fabre writes, "one knows the nature and essence of this central salt, he knows also the essence of nature, along with all its qualities, virtues, and powers." Thus alchemy led not just to a knowledge of metals, but to universal wisdom, an understanding that the creation contained both pure and impure parts, and that, by means of specific processes, it was possible to change impure portions into complete and wholly perfect things. "The original purity [*purum*] of nature" is, he wrote, "the basis and foundation of medicine." It is "the author of all light. . . and is the seat of all things mixed into nature." Through *chymia* one learned how to obtain that purity "which alone is capable of action and virtue" and from which the greatest medicaments are invented.[36]

All nature, Fabre argued, was animated, in the sense that everything possessed a Form or soul and an innate heat. Heavenly rays, which, he explained, were nothing other than subtle spirits, descended down from the heavens mixing themselves with the elements to bring forth all the lower things. Thus, contrary to Aristotle and embracing a position common among hermetic writers, Fabre erased material distinctions between the celestial and terrestrial worlds. Heavenly spirits were of the same matter as terrestrial elements. They had to be. Otherwise they would not mix and produce homogenous things. So, he declared, "it is not to be doubted that the heavens and the lower elements are of one common substance."[37] Like Nasari, Fabre also relied upon enigma and symbol to convey secret truths. The seven-headed hydra was his symbol for the process of producing the Philosophers' Stone. Each head (representing each of the metals) was, in turn, cut off and destroyed by the chymist's fire producing "a unique, homogeneous and uniform substance . . . more distinguished than ordinary gold."[38]

What affect books like these had upon the patron of Gacian is difficult to know. Lastanosa, who was called "erudite in alchemy and other arts," seems to have prized claims to practical effects over universal virtues in the spagyric medicines produced in Huesca. Nevertheless, there were rumors of great secrets collected there as well. Diego Bercebal, who visited the house of Lastanosa and found there many exquisite remedies, is reported to have been disappointed not to discover in Huesca what others had told him he might find there, namely, the Avicula of Hermes Trismagistos (a prime matter distilled from Mercury) and the universal solvent called Alcahest.[39] Although we can determine that Vincencio Juan de Lastanosa collected texts that made ample reference to universal medicines, we do not know, and perhaps can never know, what his actual experience of such medicines was. Quite likely however, for him as for many others, experience itself would have been the final judge in determining the value of the virtues such recipes promised. He might have agreed on that account, without knowing it of course, with the later Italian physician working in Spain, Giuseppe Gazola (1661-1715), that there was no better teacher, after all, than experience. And perhaps, if his experience with the claims made for elixirs and panaceas was at all disappointing, he may also have concluded regrettably with Gazola that, after all was said and done, the true universal panacea was really composed of four essential and well-known ingredients: diet and rest mixed with time and suffering.[40]

Notes

1. For works related to alchemy and chemical medicine at the house of Lastanosa see Miguel López-Pérez, *Asclepio renovado. Alquimia y medicina en la España Moderna (1500-1700)*, (Madrid, Ed. Corona Borealis, 2003), 268-284.
2. López-Pérez, *Asclepio Renovado*; Mar Rey-Bueno, *Alquimia: el gran secreto*, (Madrid, EDAF, 2002); Pascale Barthélemy, *La sedacina ou l'oeuvre au crible: l'achimie de Guillaume Sedacer, carme catalan de al fin du XIVe siècle*, (Paris, S.É.H.A.; 2002); Mar Rey-Bueno & Mª Esther Alegre-Pérez, "Los destiladores de Su Majestad: destilación, espagiria y paracelsismo en la corte de Felipe II," *Dynamis*, 21 (2001): 323-350; Jesús Osácar Flaquer, *La alquimia en Aragón*, (Zaragoza, Ibercaja, 1999).
3. *Musaeum hermeticum, omnes sopho-spagyricae artis discipulos fidelissime erudiens, quo pacto suma illa veraque medicinia. . .queat. . .* (Frankfurt, Jennisius, 1625). Enlarged edition: *Musaeum hermeticum reformatum et amplificatum . . .* (Frankfurt, apud Hermannum a Sande, 1678).
4. Jürgen Habermas, trans. Jeremy J. Shapiro, *Knowledge and human interests*, (Boston, Beacon Pr., 1972)
5. The earliest reference I have found in a brief survey is to an edition published at Hamburg in 1631.
6. Hadrianus Mynsicht, trans. John Partridge, *Thesaurus et armamentarium medico-chymicum or a treasury of physick. . .*, (London, printed for J.M. for Awnsham Churchill, 1682), 1. The Latin text I have used is *Thesaurus et armamentarium medico-chymicum . . . cui in fine adjunctum est testamentum Hadrianeum de aureo philosophorum lapide. . .*, (Venice, apud Combi et La Noù, 1696), 1.
7. Mynsicht, *Thesaurus*, 14-15; (13-14 in 1696 edition).

8. Mynsicht, *Thesaurus*, 4-5; (4 in 1696 edition).
9. Mynsicht, *Thesaurus*, 5; (5 in 1696 edition).
10. Mynsicht, *Thesaurus*, 185-186; (223 in 1696 edition).
11. Miguel López-Pérez, "La alquimia: un problema social en la España del siglo XVII", *Llull*, 23 (2000): 643-659.
12. Martin Pansa, *Pharmacotheca publica et private, das ist: Stadt- Hoff- und Hauß Apothecke. . .* , (Leipzig, In Vorlegung Henning Grossen des Jüngern S. Erben, Gedruckt bey Johann Glück, 1622-1623).
13. Diego Bercebal, *Recetario medicinal espagirico. Obra posthuma de Fr. Diego Bercebal, enfermero mayor del real convento de S. Francisco de Zaragoza. . . Sacala a Luz D. Miguel Pasqual, sindico del mismo convento*, (Zaragoza, por Luis de Cueto, impressor de la Universidad, 1734), 63-65; 68; 82; 100.
14. Mynsicht, *Thesaurus*, 68-83; (73-93 in 1696 edition).
15. Bercebal, *Recetario*, 27.
16. Mynsicht, *Thesaurus*, 169-170; (204 in 1696 edition).
17. Mynsicht, *Thesaurus*, 100; (113-114 in 1696 edition).
18. Mynsicht, *Thesaurus*, 90; (101 in 1696 edition).
19. Mynsicht, *Thesaurus*, 60-62; (65-66 in 1696 edition).
20. Mynsicht, *Thesaurus*, 41-42; (42-43 in 1696 edition).
21. Mynsicht, *Thesaurus*, 150; (177-178 in 1696 edition).
22. Mynsicht, *Thesaurus*, 54-55; (58-59 in 1696 edition).
23. Mynsicht, *Thesaurus*, 34-38; (34-39 in 1696 edition).
24. Recorded in Johann Wittich, *Bericht von den wunderbaren Bezoardischen Steinen so wieder allerley gift krafftiglich dienen*, (Leipzig, Lantzenberger, 1592, first published 1589), 1-14.
25. Frederick Slare, *Experiments and observations upon oriental and other bezoar-stones which prove them to be of no use in physick. . .*, (London, for Tim. Goodwin, 1715), 34.
26. WITTICH, *Bericht*, 5.
27. Adrian Mynsicht, *Aureum speculum redivivum. Das ist: die uhralte entwichene Güldene Zeit. . .* (Frankfurt, Jennis, 1625): Also *Musaeum hermeticum. . .*, (Frankfurt, sumptibus Lucae Jennisii, 1625).
28. First published as *Il metamorfosi metallico et humano. . .* (Brescia, 1564).
29. Giovanni Battista Nazari, *Della tramutatione metallica sogni tre. . . nel primo de quale si tratta della falsa tramutatione sofistica: nel secondo della utile tramutatione detta reale usuale: nel terzo della divina tramutatione detta reale filosofica. . .*, (Brescia, Pietro Maria Marchetti, 1599), 147-148.
30. Pierre-Jean Fabre, *Propugnaculum alchymiae. . .*, (Toulouse, apud Petrum Bosc, 1645), 4; 7.
31. Pierre-Jean Fabre, *Alchymista christianus in quo Deus rerum author et quamplurima Fidei Christianae mysteria, per analogias Chymicas et Figuras explicantur, Christianorumque Orthodoxa, doctrina, vita et probitas non oscitanter ex chymica arte demonstrantur*, Toulouse, apud Petrum Bosc, 1632).
32. Pierre-Jean Fabre, *Hercules Piochymicus. . .*, (Toulouse, apud Petrum Bosc Bibliopolam, 1634), 1-2.

33. Pierre-Jean Fabre, *Des scharffsinnigen, weltberuffenen, und unvergleichlichen Philosophi Petri Johannis Fabri. . . Alle in zwei Theile verfasste Chymische Schriften. . .*, (Hamburg, in Verlegung Lucas Eding, 1713), 92-93.
34. Fabre, *Hercules Piochymicus*, 5.
35. Fabre, *Des scharffsinnigen* , 2.
36. Pierre Jean Fabre, *Myrothecium spagyricum sive pharmacopoea chymica, occultis naturae arcanis, ex Hermeticorum Medicorum scrinijs depromptis abunde illustrata*, (Toulouse, apud Petrum Bosc, 1628), 10-13.
37. Fabre, *Des scharffsinnigen*, 86.
38. See Bernard Joly, *Rationalité de L'alchimie au XVIIe Siècle*, (Paris, Librairie Philosophique J. Vrin, 1992), 183.
39. Bercebal, *Recetario*, A5v.
40. In Flavio Ballestrasse, *Studi sul settecento medico "contro I dommatici e I galenisti" del Dott. Giuseppe Gazola (1716)*, (Pisa, Casa Editrice Giardini, 1970), 26; 47.

CHAPTER TEN

THE COLLECTOR OF SECRETS: POTABLE GOLD, AN ITALIAN ALCHEMIST AND A NURSE-SOLDIER IN LASTANOSA'S LABORATORY*

MAR REY BUENO

On November of 1657, prince Felipe Próspero, son of Felipe IV and his second wife Mariana of Austria was born in Madrid. After the death of Balthazar Carlos, the monarch's first-born child, Felipe Próspero's birth secured the continuity of the Hispanic branch of the Hapsburg dynasty, which was threatened by the absence of a male descendant. People rejoiced throughout the kingdom and celebrated the long awaited birth with parties that filled the streets of the empire with all kind of spectacles. One of these celebrations took place in Huesca. An anonymous chronicler recorded the festivities in the *Relación de las fiestas que la ciudad de Huesca de el Reyno de Aragón ha hecho al nacimiento del Principe nuestro Señor D. Felipe Próspero de Austria* (Huesca, 1658).[1] He framed the account as foretelling the abundance and prosperity that such a lucky birth would bring. Civil and ecclesiastic authorities in the town generously contributed to the events held in Huesca, but the spectacle staged by Juan Vicencio Lastanosa eclipsed all the others.

The events took place in the square on Huesca's main street flanked by the front façade of Lastanosa's palace and by the church of the Jesuits and whose other sides were closed off by diverse adornments constructed for the occasion. Lastanosa sought to reproduce in the square a miniature version of the many curiosities and marvels he had accumulated in his palace and in his lavish gardens. These replicated his collection of antiquities and his gardens and up to the mirrors and optical instruments that formed part of his collection of rarities.[2] We know well the splendid museum that Lastanosa assembled in his palace, so it is surprising that, from among all the possible aspects the anonymous chronicler could have emphasized, he chose to transmit to posterity one of the least known aspects of Lastanosa's collection and interests. This was his passion for the alchemical transmutations, personified by the presence of the Italian priest Natale Baronio, who elaborated spirits and quintessences for Lastanosa during four years:

> Sea por todos celebre testimonio el Dotor D. Natal Baronio, Sacerdote Napolitano de Ilustre familia, Vicario General del Obispado de Policarpo, Varon Insigne en Teulugia, Canones, Astrologia i Medicina, i sobre todo en los milagrosos efectos de la quimia; el qual despues de aver gozado largo tiempo las grandezas i comunicacion de Varones Eminentes, i de los mayores Señores de Venecia, Florencia Roma i toda Italia, aviendo passado a ver a España, llegò a esta Ciudad movido solamente de la fama de D. Vincencio Lastanosa, de quien honroso huesped, para dexar compuesto su mas intimo Camarin de lo mas estimable de la vida, està agora sacando varios extratos de minerales, plantas i licores, haciendo la preciosa confeccion de perlas, formando la unica i singular medicina de los Reyes del oro potable, que entre otras grandezas de nuestro Gran Monarca se muestra en la Real Botica del Escurial, de que se han visto ya aqui milagrosos efectos en necessidades extremas; i assi podrè asseverar sin arrojamiento, ni hiperbole, que es oy tan memorable Huesca por D. Vincencio Lastanosa como celebre en los siglos passados por sus valerosos hijos, que le dieron el timbre de Ciudad vencedora.[3]

It could be that Baronio's temporary stay in Huesca coincided with the dynastic festivities and this might have encouraged the chronicler to highlight this aspect of Lastanosa's universe. What is certain is that this is the first of what would become a numerous list of references associating Lastanosa with alchemy, an aspect of his life that has been systematically ignored by the historians. Lastanosa's alchemical pursuits have been categorized by historians as either the necromantic pursuits of someone interested in magic[4] or as brief heterodox vagaries that later, or never, should somehow be justified.[5] Only Miguel López-Pérez's recent work has tried to rescue Lastanosa's interest in alchemy work from neglect.[6] His study of Diego Bercebal's work, the head nurse (*enfermero mayor*) of the convent of San Francisco in Zaragoza, emphasized Lastanosa's role as s scholar of chemical medicine.[7] But, Lastanosa's alchemical interest went far beyond this as is made patently evident by references made in several manuscripts linked directly to him.

We are now aware of a Lastanosa who was also an avid collector of alchemical texts, a patron of experienced and neophyte alchemists, and a translator into Spanish of *Les elements de chymie* (Paris, 1615) of Jean

Beguin, a reference work in the field of the preparation of chemical medicines. His contemporaries described him as being "an alchemical erudite" (*erudito en la alquimia*)[8], "a fountain of health and ideas to make medical treasures... from gold and pearls" (*fuentes de salud e ideas para hacer tesoros de medicina (...) del oro y perlas*)[9] and great collector of "secrets and exquisite remedies" (*secretos, y remedios exquisitos*).[10] In light of this evidence, we must revisit Lastanosa from the perspective of history of science, rather than solely from the perspectives of art or literary history. This study will attempt to analyse his alchemical interests, framing them within the context of the type of virtuoso Lastanosa was — the prototype of a scholar interested in the most varied secrets of the nature.

***

According to the learned Belgian Samuel Quicchelberg (1529-1567) in his book *Inscriptiones vel Tituli Theatri amplissimi* (Munich, 1565) —the first work containing an exhaustive review of the different kinds of objects that should form part of a proper Renaissance kunstkammern— there were the five essential categories that a collector had to have if he wanted to be considered a serious collector: religious art and topographic maps, classic sculpture and ancient numismatics, native specimens and ethnographic objects, scientific instruments and mechanical artifices, and finally, precious stones and graphical representations.[11] Three decades later, the English philosopher Francis Bacon (1561-1626) offered his own vision. From Bacon's point of view, a gentleman's knowledge rested on four pillars: a complete library, a spacious garden, an exquisite cabinet, and a perfect laboratory[12]. Quicchelberg and Bacon were the first theorists of a particular phenomenon of Early Modern Europe, the so-called kunst-und wunderkammern or cabinets of curiosities. These collections consisted of a variety of artefacts, both *naturalia* and *artificialia*, that sought to capture the greatness of the universe or macrocosms within the limits of a cabinet that in turn became a microcosmic image of the universe. The cabinet of curiosities ranged from the sumptuous chambers of monarchs and the high nobility, who used such displays to exalt publicly their image of power, to the scholars' more modest cabinets which sought to satisfy the longing for knowing. Both the grand cabinets and the more modest ones were popular throughout Europe and helped foster a fluid traffic of objects and knowledge scarcely glimpsed in the continent until then.

Vincencio Juan de Lastanosa was one of the collectors who carried out Quicchelberg's and Bacon's vision, and in the doing so transformed his palace in Huesca into a clear example of the type of residence that a gentleman of his era and stature had to possess. Lastanosa's garden, rooms of cabinets, library and chemical laboratory had their own distinct spaces within the mansion; they formed an assemblage that says much of his owner. Although Lastanosa was not the only Spanish nobleman interested in collecting marvels and rarities, he stands out for having fashioned around him all four pillars of the Baconian collection.[13]

We know the structure and design of Lastanosa'a gardens, the vast collection his cabinets held, and the size of his library, but we know very little about his laboratory. The *Descripción de las antigüedades y jardines de Don Vincencio Juan de Lastanosa*, the most detailed and important of all sources describing the palace that survives, does not mention the laboratory.[14] It could be that about 1650 —the text's possible date— Lastanosa was not yet interested in practical alchemy and had not installed a laboratory in his palace. Another reason for the absence of the laboratory from the description could be that Juan Francisco Andrés of Uztarroz, its author and a good friend of Lastanosa, intentionally ignored this aspect, judging it perhaps not being interesting or opportune. If we ascribe to the second reason, we would be following the position generally held by historians. They interpreted Lastanosa's alchemical practice as a somewhat delicate subject, an activity Lastanosa pursued in private, having been fascinated in his youth by the alchemical dream, and that after numerous unsuccessful experiences, left him disillusioned and in need to explain to his contemporary the reasons for his heterodox inconstancies.[15] I prefer the first possibility, which suggests the figure of a younger Lastanosa as collector of books of alchemy, but who did not become involved in alchemical trials until the arrival in Huesca, about 1657, of the Neapolitan priest Natale Baronio.

There are no references that describe a young Lastanosa as a practicing alchemist, but rather that show him to be collector of alchemical books in the years prior to 1657. His library's catalogue, which might have started around 1640, had an important selection of the principal alchemical texts published in sixteenth and seventeenth centuries.[16] He had compilations such as the first five volumes of the *Theatrum Chemicum,*[17] the *Museum Hermeticum,*[18] or the *Bibliotheca Chemica Contracta.*[19] He also had works of strong symbolic character, as Giovanni Battista Nazari's *Della tramutatione metallica sogni tre,*[20] or Cesare della Riviera's *Il magico mondo de gli heroi,* [21] along with texts on metallurgy, like *De re metallica* of Georg Agricola.[22] These were joined by manuals of chemical medicine, such as Oswald Croll's *Basilica Chymica,*[23] *Les elements de chymie* of Jean Beguin,[24] or Adrian von Mynsicht's *Thesaurus et armamentarium medico-chymicum.*[25] These stood along books of secrets, such as Alessio Piemontese's *Secreti*[26] or Giambattista della Porta's *Magia Naturalis.*[27]

Lastanosa was not a mere bibliophile interested in hoarding all that was published on the topic. The analysis of the outlined works suggest a tendency towards the world of the so-called medicinal secrets, which was one of the most popular genres of the early modern era. The most famous of these books was the *Secreti* of Alessio Piemontese, who became one of the most popular and prominent figures of sixteenth century Europe. This book had more than hundred editions between 1555 and 1699.[28] Piemontese's work was in the library of Lastanosa, along with those of other teachers of secrets, and which are also described in the *Narración de lo que le pasó a Don Vincencio Lastanosa a 15 de octubre del año 1662 con un religioso docto y grave*[29], a very important document for the study of the Lastanosa's alchemical interest. It describes that Lastanosa had a good number of the books of secrets published up to 1662: "*De Secretos, Alberto. Uvecherio. Porta. Cardano. Alexos Piamontes, Fiorabanto, Fallopio. Lunomio. Filatro...*"[30]

We can augment this selection by referring to his catalogue, where we find more books of secrets written by Spanish authors. Although books of secrets were an eminently Italian genre, these tomes are worthy representatives of this movement. In Spain, we can find authors such as the Valencian polygraph Jerónimo Cortés[31], the doctor of Teruel Jerónimo Soriano[32] or the Catalan friar Miguel Agustín.[33] All of these books were kept in the fourth desk of Lastanosa's library: "*Quarto escritorio. Es de libros chimicos, debense los mas raros a la liberalidad del Ilustre Señor D. Camilo Locarni, noble veneciano.*"[34] Next to these printed books, Lastanosa also possessed at least one manuscript volume described as a compilation of numerous secrets summarized of various sources.[35]

His collection of alchemical books might be explained as just another requisite aspect within Lastanosa's general plan, which was guided by his interest in the most wonderful and rarest *naturalia*, *artificialia*, *exotica* and *mirabilia* that existed in the world. But the *Narración* of 1662 undermines this hypothesis. This document written by Lastanosa describes in detail his involvement in chemical practices. This is the document where some historians see a Lastanosa who has to somehow justify his interest in the field.[36] A careful reading of the *Narración*, however, suggests instead a Lastanosa more interested in the theoretical aspects of alchemy, that is *before* the arrival of the Neopolitan Baronio in Huesca.

Lastanosa's interest in alchemy was, in fact, more important to him than other aspects of his collections. As we can read in the *Narración*:

> "Ingenuamente responderé, Padre, que aviendo tenido grandes ocasiones de saber dessa prodigiosa arte, creiendo que el fin de ella era enriquecer el hombre, siempre lo dexe despreciando ese modo de tesoros. Hasta que llegado avrá como 4 años a mi casa el Doctor Nadal Baronio le di de mano, como he hecho con muchos otros, y el viéndose despreciado aviendo llegado a esta Ciudad solo por verme y comunicarme, me dixo: Ha Señor, que yo no soy burlador ni tramposo, que soi sacerdote y doctor teologo y medico, y mi chimica se encamina toda a la salud del hombre, a alargarle la vida y a moderar sus accidentes, y esto lo obro con el oro potable, plata potable, el espiritu, quinta esencia, sal, extractos, y magisterios de las perlas, coral, ambar, carabes, etc. Y yo si os he de hacer oro potable no os pediré oro (que esso podia hazerme sospechoso), porque yo desta tierra que pisamos saco oro"[37]

By receiving in his palace at Huesca the Neapolitan Baronio, Lastanosa has able to emulate monarchs and the nobility who became patrons and protectors of an relatively new branch of alchemy that emphasized the creation of medicines and medical therapies using elixirs and quintessences. This desire to emulate is evident by the comments made by the anonymous narrator of the *Relaccion* of 1658, when he compares the elixirs and quintessences made in the laboratory in Huesca with those prepared in the Mansion of the Waters at El Escorial. In this way, Lastanosa used his sponsorship of alchemical practices to fashion an aristocratic image of power. An image he had been constructing for quite some time and which was one of motivations for his palace, gardens and collections.

We should not suppose that Lastanosa's interest on alchemy and its therapeutic side was solely the result of self-promotion. In the *Habitación de las Musas, recreo de los doctos, asilo de los virtuosos*, a brief biography of the Lastanosa written by his youngest son and heir Vincencio Antonio,[38] we learn that Lastanosa translated into Spanish Jean Beguin's *Les elements de chymie* (Paris, 1615), an eminently practical manual destined to the education of apothecaries specializing in the production of chemical medicines. If Lastanosa chose to translate this text, it was, undoubtedly, because he considered it very important for laboratory practice and for the production of chemical medicines. We do not know when he translated the text, but it happened surely after the Baronio's stay in his palace. If we are to agree with what was reported in the *Narración* of 1662, this Neapolitan theologian was instrumental in teaching Lastanosa how to prepare chemical medicines and encouraged him to continue acquiring new knowledge:

> Pareciome consolarle con oirle y traerle por huesped a mi casa, donde en espacio de tres años obró cosas prodigiossissimas no siendo lo mas a algunos hombres tenidos por muertos, recibida la extrema unción, privados de todos los sentidos, con quatro gotas de oro potable restaurarlos, y dexarlos habiles para hablar, discurrir, recibir los sacramentos y hacer su testamento, sino que a algunos de estos los restauró hasta darles la salud y convalecer perfectamente. De estas cosas enriquecio mi casa, que las estimo mas que quanto tengo, pues con ellas mui frequentemente estoy socorriendo y mejorando la salud de mis amigos.[39]

Lastanosa, thus earned the reputation of a health-giver among his friends and neighbours. The news ran through Huesca and even to Zaragoza. In that city, a former soldier who had become a Franciscan nurse learned of it and set his sights on the Lastanosa palace in Huesca. Juan Bercebal (later fray Diego) decided to go to Huesca to learn how to make these chemical remedies. Bercebal was born in Tarazona (Zaragoza) in early seventeenth century and spent his early life acting as a soldier. Bercebal's superiors ordered him to write the *Recetario* out of fear that at his death a lifetime's worth of knowledge would be lost.[40] This posthumous work proves vital in helping us understand Lastanosa's chemical and therapeutic practices.[41]

In the letter approving the publication that precedes the text of the *Recetario*, Atilano Thomás Manente, a citizen of Tarazona and great friend of Bercebal, explains how Bercebal began in the chemical practice in Lastanosa's house,

> Ya professo en la Religion, le pusieron en el oficio de Enfermero, en que sin olvidar las eficacias de su natural, mostró con la aplicación, y el estudio, el cargo que se hazia, para cumplir más exactamente con su obligación. Todo era efecto de su caridad, y por satisfacerla más bien, començó a comunicar, y consultar remedios con aquellos Maestros, y Personas, que le podían instruir. Tomó su primer principio en la ciudad de Huesca, en la casa de los señores Lastanosas, donde se introduxo por este fin, por ser tan notoria la afición de aquellos señores a secretos, y remedios exquisitos, que alcanzaron con la experiencia, y comunicación de los primeros Hombres en la Química, y Medicina, assi Extrangeros, como Naturales.[42]

Bercebal and Manente would later learn from each other. After Bercebal's apprenticeship at Lastanosa's house, he went back to his home town and shared with Manente the knowledge he had recently acquired in Huesca. In return, Manente taught him own recipes and techniques. Bercebal's stay in Tarazona was not as long as he would have wanted.

Once he had become an experienced alchemical and pharmaceutical practitioner, he received a call from the Convent of San Francisco at Zaragoza to become its head nurse. Bercebal continued his education in the convent's infirmary, carrying out his own experiments and remained in communication with other experts in chemistry,

> Prosiguió allí con mas oportunidad con sus exercicios, consultando con Hombres Doctos y peritos en estas Facultades, y logró verdaderamente la Ciencia, y la experiencia, que no han alcanzado, ni logrado muchos que se tienen por Maestros de esta Profession.[43]

What did Bercebal learn while at Lastanosa's palace? We must suppose he learned a great deal given he had access to the best books on chemical medicine, as well as the knowledge that Lastanosa learned from Nadal Baronio. But principally, Bercebal learned how to make the "Sprit" he described in the first eight chapters of his *Recetario*. ("*Allí [en Huesca] le revelaron el Espiritu que pone en las ocho secciones del capítulo primero de este Recetario; bien que aquellas noticias las amplió después mucho más.*")[44] What "Sprit" is this? In the book's first chapter, titled "Of a medicinal treasure," Bercebal describes it as a universal solvent with the name of "Spirit of the Elements of the Anima Mundi" or Water of Life or Alkahest,

> Ansia ha sido de los Artífices un dissolvente Universal, con el nombre de Espíritu de los elementos de Anima Mundi, de Agua de Vida, de Alcaest & y lo buscan en varios medios, y por distintos caminos; hallando en los Autores hacia su fin, mucho de misterio y mucho mas de encanto.[45]

It is not until the sixth section of the first chapter when Bercebal mentioned that he received this knowledge from others, but he does not specify of from whom,

> Dióseme la primera noticia de esta medicina con el nombre de 'Espiritu Universal'. Yo estimé la dádiva como cosa singular; y en estos términos entiendo averle proporcionado el aprecio.[46]

We must wait until further in the text, when Bercebal finally discusses a recipe he learned from the alchemist Baronio:

> De un Don Nadal Baronio Veneciano, que fue quien dexó el dissolvente de el oro a Don Vincencio Lastanosa, tengo la preparación siguiente (…). En casa de Lastanosa he alcanzado algo de este laudano, hecho por mano de dicho don Nadal, y le vi obrar bien. Pero no satisface una obra prolixa, costosa, en que entra sin purificación el opio, y donde se juntan tantos cordiales opuestos, unos volátiles y aromáticos, otros fixos y lapideos, que

> han de embarazarse; solo la práctica común de las confecciones puede responder, como exemplar de otra escuela, sin satisfacer a los primores Químicos.[47]

The "medicinal treasure" learned in Huesca was, according to Bercebal, the most valuable knowledge he possessed,

> Tengo dedicadas todas mis tareas de enfermería a San Diego mi especial Patrono: me es de gran consuelo el esperar, que Dios se sirva de esta obra ruda. Deseo concluir en día de este glorioso Santo este especial capítulo, en que dexo manifestado lo mas util y secreto que tengo, y por lo qual me he puesto a escribir, aunque ya empeñado, me dilato otras medicinas utiles.[48]

Bercebal's words are extremely important; they reveal the real nature of the alchemical practices that took place in Lastanosa's palace. As Bercebal declared, he learned how to prepare the so-called universal dissolver, one of the most sought after alchemical agents, in Lastanosas's laboratory. But Bercebal's words suggest something more. It also suggests that Lastanosa was a follower of the iatrochemical theories of Jean Baptiste van Helmont (1579-1644).

In his *Ortus medicinae* (Amsterdam, 1648), van Helmont thanked God for calling him to the ways of the art of fire, a metaphorical reference to chemical arts, because it was through the art of fire that van Helmont thought it possible to attain the greatest secret of the nature: the seminal virtue from which all the things originated.[49] Van Helmont postulated that the intimate essence of all matter was in the spirit obtained via the volatilization of the salt. This was the final product of the diverse operations carried out by the chemists that sought to obtain the purest material substance, the origin of all creation. This, and no other, was the universal spirit Bercebal learned to prepare in Lastanosa's house and whose purpose was to dissolve gold into the three elemental substances: pure sulphur, pure mercury, and pure salt.[50] As opposed to the alchemists engaged in the art of making gold, the medical-chemists were interested on how to dissolve gold, a work that they saw as simpler than that of the alchemists. The medical-chemists used the same techniques and processes of the alchemist to 'open' gold to capture its essential components that they saw as a key capable of opening the most coveted casket. It was a task they perceived as easier than elaborating gold from the corrupt matter.[51]

We return, finally, to an apparently insignificant comment made by Lastanosa in the *Narración* of 1662, but that now seems highly suggestive. It is the response that Lastanosa gave to the learned and serious monk about why he is not discussing his alchemical interests, and begins with the preamble, "*Ingenuamente responderé, Padre, que aviendo tenido grandes ocasiones de saber dessa prodigiosa arte, creiendo que el fin de ella era enriquecer el hombre, siempre lo dexe despreciando ese modo de tesoros.*" The text continues with an account of Baronio's arrival to his palace and the chemical trials carried out by the cleric over the following four years. This preamble refers to his beliefs *before* 1657, when Lastanosa thought that the sole purpose of alchemy was obtaining the philosophical stone; after this date Lastanosa focused on alchemy and alchemical practices but from a new perspective, thanks to the knowledge that he received from the monk Baronio. This change of perspective locates Lastanosa next to many European scholars who used chemical philosophy as a means of approaching the study of nature and for the purpose of discovering its mysteries. What Lastanosa sought was not simply to hoard and display rarities and marvels. What Lastanosa sought was to accumulate knowledge in order to read the book of the nature as the maximum expression of the Divine.

***

I will conclude here this study of Lastanosa's alchemical interests. There still remains much work to be done on the topic, from a rigorous analysis of the treatises in his library to the search of new references that might help us outline better this new facet of Lastanosa. New sources might confirm or invalidate the conclusions raised here. But what is certain is that as a collector of secrets Vincencio Juan Lastanosa had a life long and genuine interest in several aspects of alchemy.

Notes

* I would like to acknowledge María M. Portuondo for their helpful suggestions in the english translation; Carlos Garcés Manau, member of the Instituto de Estudios Altoaragoneses, for providing me copies of all primary sources relating to Lastanosa; and José Rodríguez Guerrero, publisher of Azogue Journal, for supplying information about the alchemical books in Lastanosa's library.

1. Biblioteca Nacional de Madrid, VE-63/40.

2. Alberto del Río Nogueras, "Literatura y fiestas en la Huesca del Siglo de Oro", in: *Signos. Arte y Cultura en Huesca. De Forment a Lastanosa. Siglos XVI-XVII*, (Huesca, Diputación, 1994), 144-151.

3. *Relación de las fiestas que la ciudad de Huesca de el Reyno de Aragón ha hecho al nacimiento del Principe nuestro Señor D. Felipe Próspero de Austria*, (Huesca, 1658), 7-8.

4. Santiago Echandi Ercila, "Filosofía y saberes del Barroco (en las bibliotecas de Spinoza y de Lastanosa)", in: José Enrique Gil Laplana, ed., *La cultura del Barroco. Los jardines: arquitectura, simbolismo y literatura (Actas del I y II Congreso en torno a Lastanosa)*, (Huesca, IEA 2000), 43-80 (50-52).
5. José Miguel Morán Turina, "Los prodigios de Lastanosa y la habitación de las musas. Coleccionismo ético y coleccionismo ecléctico en el siglo XVII", *Separata*, 5-6 (1981): 53-59 (55). Fermín Gil Encabo, "Vincencio Juan de Lastanosa y sus prodigios", in: *Signos. Arte y Cultura en Huesca. De Forment a Lastanosa. Siglos XVI-XVII*, (Huesca, Diputación, 1994), 111-123 (120-121).
6. Miguel López-Pérez, *Asclepio renovado. Alquimia y medicina en la España Moderna (1500-1700)*, (Madrid, Ed. Corona Borealis, 2003), 268-284; Miguel López-Pérez, "Anatomía del virtuoso: coleccionismo y melancolía en la figura de Vincencio Juan de Lastanosa", *Argensola*, 115 (2005): 167-185; Miguel López-Pérez, "La alquimia y Vincencio Juan de Lastanosa", in: *Vincencio Juan de Lastanosa (1607-1681). La pasión de saber*, (Huesca, Instituto de Estudios Altoaragoneses, 2007), 177-183;
7. Diego Bercebal, *Recetario medicinal espagírico*, (Zaragoza, Diego Larumbe, 1713).
8. Diego Vincencio Vidania in his introduction to *Tratado de la moneda iaquesa, y de otras de oro, y plata del Reyno de Aragón* (Zaragoza, 1681), by Vincencio Juan de Lastanosa.
9. José Alfay, in his introduction to *Sarao de Aranjuez de varios versos y novelas* (Madrid, 1666), by Jacinto de Ayala.
10. Atilano Tomás Manente in his introduction to *Recetario medicinal espagírico* (Zaragoza, 1713), by Diego Bercebal.
11. Samuel Quicchelberg, *Inscriptiones vel Tituli Theatri amplissimi, complectentis rerum universitatis singulas materias et imagines eximias, ut idem recte quoque dici possit: Promptuarium artificiosarum miraculosarumque rerum, ac omnis rari thesauri et pretiosae supellectilis, structurae atque picturae, qual hic simularis theatro conquiri consuluntur, ut eorum frequenti inspectione tractationeque, singularis aliqua rerum cognitio et prudentia admiranda, cito, facile ac tuto comparari possit*, (Monachii, Ex officina Adami Berg typographii, 1565).
12. In *Gesta Grayorum* (1594), we can read: "*First, the collecting of a most perfect and general library, wherein whosoever the wit of man hath heretofore committed to books of worth... may be made contributory to your wisdom. Next, a spacious, wonderful garden, wherein whatsoever plant the sun of divers climate, or the earth out of divers mouls, either wild or by the culture of man brought forth, may be... set and cherished: this garden to be built about with rooms to stable in all rare beasts and to cage in all rare birds; with two lakes adjoining, the one fo fresh water the other of salt, for like variety of fishes. And so you may have in small compass a model of the universal nature made private. The third, a goodly, huge cabinet, wherein whatsoever singularity, chance, and the shuffle of things hath produced; whatsoever Nature has wrought in things that want life and may be kept; shall be sorted and included. The fourth such a still-house, so furnished eith mills, instruments, furnaces, and vessels as may be a palace fit for a philosopher's*

stone". Oliver Impey & Arthur MacGregor, *The origins of museums: the cabinet of curiosities in sixteenth and seventeenth centuries Europe*, (Oxford, Clarendon Press, 1985), 1.
13. For a long list of Wunderkammern in 17th century Spain, see: José Miguel Morán Turina & Fernando Checa Cremades, *El coleccionismo en España. De la cámara de maravillas a la galería de pinturas*, (Madrid, Cátedra, 1985).
14. Hispanic Society of America (HSA), Ms. B-2424, ff. 24r-51v°.
15. This is what Fermín Gil Encabo concludes in Gil Encabo, "Vincencio Juan de Lastanosa y sus prodigios", to justify Lastanosa's alchemical activities. In addition he maintains that King Philip II fomented similar activities. This suggests that Philip II also hid his passion for alchemy. There is nothing further from reality. Neither of them, especially given their privileged social positions, needed to justify their actions. Alchemy was not frowned upon during the early modern era. Philip II's interest in using alchemical practices to obtain a universal remedy was the reason that he established in El Escorial the most impressive laboratory of its time and unlike any other in all of Europe. For more on Philip II's alchemical interest, see Mar Rey-Bueno, *Los señores del fuego. Destiladores y espagíricos en la corte de los Austrias*, (Madrid, Ed. Corona Borealis, 2002).
16. Stokholm Royal Library, *Sparvenfeldt* Catalogue (K. B. Sp. 10-U 379), entitled *Catálogo de los libros. D. Vincencio Ioan Lastanosa. Por orden de alfabeto*. This manuscript was partially printed by Karl-Ludwig Selig, *The Library of Vincencio Juan de Lastanosa, Patron of Gracián*, (Genevae, 1960). In addition: Ricardo del Arco y Garay, *La erudición aragonesa del siglo XVII en torno a Lastanosa*, (Madrid, 1934), 199-215.
17. *Theatrum chemicum, præcipuos selectorum auctorum tractatus de chemiæ et lapidis philosophici antiquitate, veritate, iure, præstantia et operationibus, continens...*, (Strassburg, Zetzner, 6 vols., 1602-1661).
18. *Musæum Hermeticum, omnes sopho-spagyricæ artis discipulos fidelissime erudiens, quo pacto summa illa veraque Medicina, qua res omne, qualemcumque defectum patientes, instaurari possunt (quæ alias Benedictus Lapis Sapientum appellatur) inveniri ac haberi queat inveniri ac haberi queat. Continens tractatus chymicos novem præatantissimos, quorum nomina et seriem versa pagella indicabit. In gratiam filiorum doctrinæ, quibus Germanicum Idioma ignotum, in Latinum conversum ac juris publici factum*, (Frankfurt, Lucas Jennis, 1625).
19. Nathan Albineus, *Bibliotheca chemica contracta ex delectu & emendatione Nathanis Albinei…*, (Genevae, Johannes Ant. & Samuel de Tournes, 1653).
20. Giovanni Battista Nazari, *Della tramutatione metallica sogni tre, di Gio. Battista Nazari bresciano; nel primo de quali si tratta della falsa tramitatione sofistica, nel secondo della utile tramutatione detta reale usuale, nel terzo della divina tramutatione detta reale filosofica. Aggiontoni di nuovo la concordanza de filosofi & loro prattica; … Con un copioso indice per ciascun sogno de gli auttori & delle 'opere c'hanno sopra di ciò trattato*, (Brescia, appresso Pietro Maria Marchetti, 1599). Published originally as, *Il metamorfosi metallico et humano, di Gio. Bat. Na. bre. nel quale si contengono quattro sogni, il primo de'quali e della tramutatione sofistica de' metalli, il secondo della tramutatione fisica de corpi*

humani; et il quarto della tramutatione spirituale in Christo, (Brescia, a instanza di Francesco Marchetti al Segno dell'Ancora, 1564).
21. Cesare della Riviera, *Il magico mondo de gli heroi...*, (Mantua, Francesco Ossana, 1603).
22. George Agricola, *De re metallica libri XII. Quibus officia, instrumenta, machinae, ac omnia denique ad metallicam spectantia, non modo luculentissime describuntur, sed et per effigies ... adjunctis latinis germanicisque appellationibu ita ob oculos ponuntur, ut clarius tradi non possint. Ejusdem de animantibus subterraneis liber ... Cum indicibus diversis*, (Basle, H. Froben & N. Bischoff).
23. Oswald Croll, *Basilica Chymica, continens philosophicam propria laborum experientia confirmatam descriptionem et usum remediorum chymicorum selectissimorum e lumine gratiae et naturae desumptorum. In fine libri additus est autoris ejusdem. Tractatus novus de signaturis rerum internis...*, (Francofurti, impensis Godefridi Tampachii, 1608).
24. Jean Beguin, *Les elemens de chymie*, (Paris, 1615).
25. Adrian von Mynsicht, *Thesaurus et armamentarium medico-chymicum: in quo selectissimorum contra quosuis morbus pharmacorum conficiendorum secretissima ratio aperitur, una cum eorumdem virtute, usu & dosi. Cui infine adjunctum Testamentum Hadrianeum de aureo philosophorum lapide*, (Lübeck, Valentin Schalhertz for Heinrich Scheruwebel, 1638).
26. *Secreti del reverendo donno Alessio Piemontese. Nuovamente posti in luce. Opera utile, et necessaria universalmente à ciascuno*, (Venetia, per Sigismondo Bordogna, 1555).
27. Giovan Battista Della Porta, *Magiae Naturalis Libri XX*, (Neapoli, apud Horatium Salvianum, 1589).
28. William Eamon, *Science and the Secrets of Nature. Books of Secrets in Medieval and Early Modern Culture*, (Princeton, Princeton University Press, 1994).
29. HSA, Ms. B-2424, ff. 52r-79v°.
30. HSA, Ms. B-2424, ff. 59v°-60r. the cited authors are, respectively, Alberto Magno, Johann Jakob Wecker, Giambattista della Porta, Girolamo Cardano, Alessio Piemontese, Leonardo Fioravanti, Gabrielle Fallopio, Lievin Lemmens and Konrad Gesner.
31. Jerónimo Cortés, *Libro de phisonomía natural y varios secretos de naturaleza*, (Valencia, Juan Crisóstomo Gárriz, 1597).
32. Jerónimo Soriano, *Libro de experimentos médicos fáciles y verdaderos recopilados de gravísimos autores*, (Zaragoza, Juan Pérez de Valdivielso, 1598).
33. Miguel Agustín, *Llibre dels secrets de agricultura, casa rustica y pastoril recopilat de diuersos autors, antichs y moderns, de llenguas llatina, italiana y francesa, en nostra vulgar llengua catalana*, (Barcelona, en la estampa de Esteve Liberôs, 1617).
34. HSA. Ms, B-2424, fol. 72r. I could not find any reference to the Locarni mentioned in the text and supplier of more relevant chemical texts in Lastanosa's collection.
35. The library catalogue, under the heading of "Section of Manuscripts and other rare papers," there is a mention of a "*Secretos en 8° manuscritos*". This reference

also appears in the Section *"Secretos varios manuscritos i impresos en octabo"*, but we cannot establish if the refer to the same manuscript.
36. *"Aunque despreciadas ya por la ciencia académica seguían teniendo un cierto peso residual entre los subproductos de la ciencia española de este momento"* ("Although already despised by academic science they continued to have certain residual height in the Spanish science of the time.") Morán, "Los prodigios de Lastanosa", 55. One more time, it is necessary to highlight how unacquainted some have been of the historiography of alchemy. The works of Allen G. Debus, Frances Yates, Betty Jo Teeter Dobbs, William Eamon, Sylvain Matton, Bruce T. Moran, William R. Newman, Pamela H. Smith, or Lawrence Principe, for example, make it impossible to maintain positions like the one noted above.
37. HSA. Ms, B-2424, fol. 79r.
38. HSA, Ms. B-2424, ff. 1r-5vº.
39. HSA, Ms. B-2424, ff. 79r-79vº.
40. *"Este fue el motivo de mandarle escribir este Recetario Químico, quando ya muy quebrado de salud, porque no se sepultasen con el sus grandes estudios, trabajos y experiencias, que en beneficio de los Enfermos había adquirido en tantos años de Enfermero. Poco despues de aver concluido esta pequeña obra, con el animo de dexarla assi manuscrita, como Alaja de la Enfermeria, para el uso de los Enfermeros curiosos, y aplicados, murió con mucho exemplo en el año de 1707, en el mes de octubre"*. Aprobación de fray José Antonio de Herrera.
41. *Recetario medicinal espagírico. Obra posthuma de Fr. Diego Bercebal, Enfermero Mayor del Real Convento de S. Francisco de Zaragoza. Dedicada por el mismo autor a los Enfermos y encomendada a los religiosos enfermeros. Sácala a la luz D. Miguel Pascual, Síndico del mismo convento*, Zaragoza, Diego de Larumbe, 1713. I consulted the text at the Zaragoza City Library-Archive, sign. A-596. There was another printed edition made in 1734 in Zaragoza by Luis de Cueto.
42. *Notas, o advertencias a la presente obra, hechas con singular reflexión, por el Inteligentissimo Maestro D. Atilano Thomas Manente, Ciudadano de Tarazona.*
43. Ibidem.
44. Ibidem
45. Bercebal, *Recetario*, 2-3.
46. Bercebal, *Recetario*, 16.
47. Bercebal, *Recetario*, 64-65.
48. Bercebal, *Recetario*, 30-31.
49. Jean Baptiste Van Helmont, *Ortus Medicinae. Id est, Initia Physicae Inaudita. Progressus medicinae novus, in Morborum Ultionem ad Vitam Longam*, (Amsterdam, Ludovicus Elzevir, 1648), 463.
50. According to Bercebal in his *Recetario*. I thank to Miguel López-Pérez, who, in a few words, helped me understand this important theory that so preoccupied chemical philosophers for more than a century.
51. On medical-chemists and chemical-physicians, see López-Pérez, *Asclepio Renovado*. Bercebal wrote his *Recetario* during the early 18th century, when van Helmont's theory were falling out of favor. This is why Manente, in his *Notas* confessed: *"Ya se desengañó [Bercebal] no ser lo que en Huesca le dixeron ser: el*

Anima mundi, la Avicula de Hermes Trimegisto, el Alchaest, la Agua de la Vida de Alderete, el Dissolvente universal, y tantos nombres como le han puesto diferentes Autores, assi Antiguos como Modernos, sino ser un Espiritu excelente a muchas enfermedades, y dissolvente a sus materias homogéneas. Siempre he dudado lo aya conseguido alguno, pues tendría una Medicina universal; y lo he tenido siempre como primera materia del Philosopho".

CHAPTER ELEVEN

LEONHART FUCHS IN THE LIBRARY AND GARDEN OF VINCENCIO JUAN DE LASTANOSA: MAIZE, CHILE, NARCISSI AND TULIPS

RAFAEL CHABRÁN

"Siguen de Flora las hermosas huellas
Las flores, que pudieran ser estrellas
errantes por lo vario i por lo hermoso,
aunque en la duración no lo parecen,
porque más que otras plantas permanence."

['And then follow from the Godess Flora,
her beautiful footprints, the flowers
which could be wondering stars
because of their variety and the beauty,
because more than other plants, they are everlasting.']

—Juan Francisco Andrés de Uztarroz[1]

"May this volume serve as the menu card of reason at the feast of the intellectuals, and thus as a register of the choice morsels being served in all the other works, to the end that their genial wisdom may be enjoyed more widely."
—Lastanosa[2]

Introduction

The attempt of putting all of nature, or rather all knowledge concerning nature into one space begins, perhaps, with Pliny's *Natural History*, the most famous encyclopedia of natural history of Antiquity.[3] As we know, that natural history was an inventory or enormous archive and description and accumulation of facts concerning nature, especially, information dealing with minerals, vegetables and animals.[4] A good argument could be made that with Pliny's work the history of natural history begins.[5] His work was also the foundation on which the subsequent history of museums and collections of curiosities begins. As has been observed, these museums were also called studios, theaters, and cabinets of curiosities, galleries, cabinets of wonders, archives, treasures, *pandechiones*, *kunstkammer* or *wunderkammer* during the Sixteenth and Seventeenth centuries.[6]

For some collecting is hobby which consists of the mere collection and organization of things or objects of the same class or objects of different types which are brought together by someone according to individual taste or for mere curiosity. Sometimes these collections are ordered, organized and even catalogued. Other times they represent a mere hodgepodge of objects, miscellanea, which seem to be collected with no visible criteria and thus being the products of obsessive behavior. During the Renaissance and Baroque periods there were famous museums and cabinets of curiosities such as those of Ferrante Imperato, Ulisse Aldrovandi, Giovan Battista della la Porta, Francesco Calzolari, Athanasius Kircher, and Ole Worm to mention but a few.[7] In Spain we also find examples of these types of museums and collection.[8] One of the most notable in that of Don Vincencio Juan de Lastanosa, tireless collector, antiquarian, virtuoso, and one of the most famous native sons of the city of Huesca, in the Spanish region of Aragon. Lastanosa has been called "One of the most interesting and exceptional collectors of Spain during the seventeenth century."[9] Among the many things that Lastanosa collected we find the works of the noted German botanist Leonhart Fuchs.

Apropos of an Approach

To date, our research into the life and work of Lastanosa indicates that he is not mentioned in any work or study of Spanish botany or horticulture, even though his library contained some of the most imortant works of botany and horticulture of his time. It is the primary purpose of the present study to examine the importance of the work of Leonhart Fuchs whose works were found in Lastanosa's library and perhaps in his famous garden. As far was we know, Fuch's works as the are found in his library have never been the object of detailed or critical study. In order to approach the study of Fuchs in Lastanosa'a library we have selected a series of framing critical questions. Among these are: (1) Of what importance where the works of Fuchs to Lastanosa?; (2) Why did

Lastanosa have the works of a famous Protestant botanist in his library?; (3) To what degree could botanical knowledge from Fuchs influence the content and organization of Lastanosa'a garden in Huesca?; (4) What can be said about Lastanosa's botanical knowledge given the many botanical works found in his personal library?; (5) What specific botanical texts did Lastanosa possess and of what importance were they to the history of botany in Spain?; and, lastly, (6) What knowledge did Lastanosa have concerning exotic plants as described in the works of Fuchs?

Furthermore, we believe that those who elect to study the life and work of Lastanosa, the noted collector and antiquarian, must do so with the necessary tools provided by the disciplines of the history of literature, history of art and historty of science. We must also recall that both during the Renaissance and Baroque periods it was impossible to draw a clear line of distinction between humanists and scientists.

Art and Science in Lastanosa

We have chosen the metaphor of "The Book of Nature" to frame the theme of "Art and Science in the Life and Work of Lastanosa." We believe thatin this way we can better study the botanical works in Lastanosa's library and his famous garden. As we know, for Lastanosa, as well as for many during the Baroque period, and before in the Renissance, art and science were intimately realated. It was impossible to separate them. Knowledge and the representation of knowledge were one. This can clearly be seen in the life and work of the Italian antiquarian Cassiano dal Pozzo and the members of the Accademia dei Lincei, whose work have many points of contact with Lastanosa. Cassiano and the members of the Accademia dei Lincei were great collectors and they were also very interested in botanical knowledge especially pertaining to plants and materia medica from the New World. For this reson we believe that scholars should approach the study of Lastanosa from an interdisciplinary, multidisciplinary or "transdisciplinary" perspective.

In order to better approach the theme of "Art and Science" with respect to flowers and plants in the life and works of Lastanosa it is instructive to review the work of the Italian Jesuit Giovanni Battista Ferrari (1583-1655) whose works are found in the Lastanosa library.[10] Ferrari was professor of Hebrew and Rhetoric at the Collegio Romano as well as being in charge of the Vatican gardens. He was also a close friend of Cassiano dal Pozzo[11], the personal secretary of Cardinal Francesco Barberni, nephew to the Pope Urban VII and a fellow Lincei and a collector in his own right. Barberini and Cassiano were also very interested in Mexican plants. Ferrari was also an influential member of the court of Urban VIII. As organizer and administrator of the Barberini Gardens he was in charge of collecting exotic plants from Asia, Africa and the Americas. Among his most important writings we can cite *De Florum Cultura* (1633).[12] This work "On Floriculture" is one of the first scholarly studies on floriculture, the horticulture science of plant farming, the cultivation of flowering and ornamental plants for the specific purpose of gardening. Ferrari's work is also a significant work in the area of chalcography, especially in terms of drawing and using copper engraving for the illustration of plants. Ferrari's book is also important because it narrates and documents the birth and origins of flower gardens as distinct from it precursor, the botanical garden or *hortus medicus* (Medicinal Herb Garden). His work is made up of four books. The first book deals with the organization of gardens. The second book deals with the cultivation of specific plants such as lilies, narcissi and carnations, some of the very plants found in Lastanosa's garden. The third book concerns plant farming as well as the connections between plants and animals. This book also deals with the aesthetic value of plants and their beauty. At the very end of this book we also find an interesting chapter on plants from the New World, as found in the Barberini gardens in Rome. We believe that this chapter is of special interest because it provides us with a snapshot of what American plants might be found in a European garden. We might even ask ourselves: Is this what Lastanosa's garden might have looked like? Ferrari's work ends with a chapter or book on flower arrangement and the production of artistic flower planters, flowerpots or vases. We should also note that Ferrari's work also contains references to Dr. Francisco Hernández (c. 1517-1587). It also contains information on tulips and the botanical works of Clusius and Dodoens, whose works were also found in the Lastanosa library.[13]

Art and science were intimately linked in the life of Lastanosa. His collecting activities often brought together his interests in both art and science. His cabinets were full of scientific instruments as well as with books on all the sciences. He also collected works of art and treatises on aesthetics. He owned treatises on art and artists such as those by Dürer, Vasari, Da Vinci and Alberti.[1415] In addition, he owned originals and copies of the works of Rubens, Jusepe Ribera and Guido Reni (1575-1642). We cannot leave the mention of Rubens without commenting on one of his works even if it is only in passing. One of Rubens masterpieces is no doubt, *The Four Philosophers* (1611-1612). In this painting we see the portraits of Rubens's friends including the lawyer Jahannes Woverius (1576-1638), the noted Dutch neo-stoic philosopher Justus Lipsius, Rubens's dead brother Phillip (1547-1611), and a self-portrait of the artist himself. We will only focus on two details.[16] First we must recall that in addition to admiring the works of Rubens, Lastanosa also owned the works of the Flemish philosopher Justus Lipsius (1547-1606).[17] Secondly, we must notice the drooping tulips next to the bust of Seneca. The reference again is to the transitory nature of life. We don't know if Lastanosa knew Rubens painting but there is no doubt that the work gives us a representation of how Lastanosa might have viewed the members of his circle.

We will end this section by concentrating briefly on another painter, Jusepe Martinez (1600-1682), one of the most important painters from the region of Aragon, and a friend of the Spanish painter Diego de Velasquez. Martinez was also a close friend of Lastanosa while being born in Zaragoza, also studied and worked in Rome where he met the Italian painter Guido Reni, of whom we will soon speak. Among Martinez most important works we have the engravings of the Life of St. Peter of Nolasco (c. 1625), which we carried out by the Dutch engraver Jahann Friedrich Greuter, based on drawings by Martinez.

Further examples of Lastanosa's interests in art and science, especially botany, are clearly seen in the work of Giovanni Battista Ferrari, *De Florum Cultura* a book found in his library.[18] In the section of Ferrari's work dedicated to the Barberini Gardens we find an engraving by Guido Reni.[19] This engraving is called "The Indies present Neptune with seeds from Indian plants so that they might be taken to the Barberini Gardens."[20] We must also point out that Ferrari's work also contains illustrations of various exotic plants, which were supposedly from the Americas. The engravings of these plants were done by the Dutch engraver Cornelis Bloemaert (1603-1692). Among these engravings we find several reproductions of flowers, which are called "Narcissi". According to some, these are representations of *Amaryllis belladonna*, also called "The Belladonna Lily." This plant is often confused, as in the work of Ferrari, with a tropical planted from the Americas related to the southern African Amaryllis of the genus, *Hippeastrum*. Before leaving this section of our essay a few more words must be said about the Dutch engraver Jahann Friedrich Greuter, mentioned above. Greuter was part of the Barberini circle and his name is closely related to the Lincei edition of the work of Francisco Hernández. Greuter was well known for his illustrations of flora and fauna in Italian works of natural history. Among these is the Accadenia dei Lincei's famous edition of the work of Hernandez, *RERUM MEDICARUM/NOVAEHISPANIAE/THEAUSRUS/SEU/PLANTARUMAN IMALIUM/MINERALIUM MEXICANORUM/HISTORIA/EX FRANCISCI HERNANDI/Novi obris Medici Primarij relationibus/in ipsa MEXICANA*

URBE conscriptis/NARDO ANTONIO RECCHO...(Romae. 1651). Both Jahann Friedrich Greuter and his father Matthäus Greuter (1564-1638) worked on the Baroque frontispiece of the Hernandez edition published by the Lincei.[21]

Leonhart Fuchs: Life and Works

Leonhart Fuchs (1501-1566), born in Wemding (Bavaria) was one of the most important physician-botanists of the 16th century. He is remembered often as one of the so-called "German Fathers of botany,"[22] which also include Hieronymus Bock and Otto Brunfels. He is perhaps best known for producing *De historia stirpium commentarii insignes …* Basel, 1542, a guide to collecting medical plants, which is considered a landmark in history of natural observation. In 1519 Fuchs went to Ingolstadt where he studied Greek and Latin and then later obtained his medical degree in 1524. Early in his life he was attracted to Protestantism. He dedicated the greater part of his life to the study and teaching of medicine, botany and pharmacology at he historic University of Tübingen, where he was a professor until his death in 1566.[23] Fuchs work, *De historia stirpium commentarii insignes …* published in Basel in 1542 is considered to be one of the masterpieces of Renaissance botany. In this work, Fuchs catalogues and describes some 400 plants from Germany, Austria as well as some plants from the New World.

Fuchs in Spain: The Case of Sevet

Even though he was a German and a Lutheran, Fuchs's work was known in Spain and translated into Spanish. Among the Spaniards who knew his work we have Miguel Servet (c. 1511-1553), the noted physician, philosopher and theologian. As we will soon see the names of Servet and Lastanosa were to be linked forever. Miguel Servet, also known as Servetus or Miguel Villanueva was born in the Aragonese town of Villanueva de Sijensa in the province of Huesca around 1511. Servet was a famous Renaissance pilgrim who traveled and studies in many places in Europe. He studied in the historic universities of Toulouse, Paris, Lyon, and Basel. During his life he wrote important works on astrology, geography, theology and medicine. He was a close friend and foe of both Catholics and Protestants. In 1553 he was arrested and accused of heresy and burned at the stake by Calvinists in Geneva after having been persecuted and hunted down by the Inquisition for many years. Servet, like Lastanosa later, was interested in the works of Fuchs. During his stay in Lyon, Servet met and developed a close friendship with the French physician Symphorien Champier (1471-1538), one of the most important humanists of the Reanissance in France. We should also remember that Champier was also one of the most important editors of the works of Arnau de Vilanova, one of the major figures in the history of science in Spain. Servet studied with Champier and the later inspired him to continue with his medical studies. Here in Huesca, the names of Miguel Servet and Lastanosa are closely connected. This connection is embodied in a famous garden and park. It is said that the city of Huesca is "a park surrounded by buildings." That park is, of course, The Park Miguel Servet, also often referred to as "the lung of the City of Hueca." He who knows Huesca, knows the Park Miguel Servet. For a long time the people of Huesca have come together in the park at the Park Bar, very close to the monument to Miguel Servet to enjoy their wine and beer. As it stands today The Miguel Servet Park dates from the time of the Spanish Second Republic in the 1930s. The park also serves as a home to the "Monument of the Small Birds," a sculpture by Ramón Acín (1888-1936), a militant aragonese anarchist who was also a noted artist and painter of the Vanguard Movement. Finally, we must remember the current Park Servet was established on part of the grounds that served as Lastanosa's famous gardens.

Fuchs in Spanish: The Translation of Juan de Jarava

Returning to Fuchs and his appearance in Spanish we must review the life and works of Juan de Jarava. Juan de Jarava, also referred to as Jaraba or Xarava, live during the Sixteenth Century and was a well-known physician, philosopher and translator. It seems that he was the personal physician to Leonor of Austria, Queen of Spain and sister of Charles V. Jarava translated the Bible, as well as the works of Cicero, Aristotle, Pliny and Erasmus. He also translated Fuchs. This translation was called *Historia de yervas, y plantas, de Leonardo Fuchsio Aleman, docto varon en Medicina, con los nombres Griegos, Latinos, y Españoles. Traduzidos nuevamente en Español con sus virtudes y propriedades, y el uso dellas, y juntamente con sus figuras pintadas al vivo ('History of Herbs, Plants by the German Leonhart Fuchs, noted doctor in Medicine, with plant names in Greek, Latin and Spanish, along with their Virtues, Properties and Uses, together with Illustrations from life').* This illustration was published in Antwerp in 1557 by the family of Arnold Birchmann in the shop of Jan de Laet. This translation was not based on the original but rather on a French translation. There are two version sof the Spanish translation. In reality it is one and the same translation with some small changes in the title pages. In the first version the title page carries the name of Fuchs as well as the translator "Iuan Iarava". These are omitted in the second version. There are two excellent studies of these edition. The first is the study by Meyers (1999), which is apart of a excellent edition and English translation of Fuchs work. The second is a superior and more complete study of the translation done by López Piñero and López Terrada (1994). In the later study, the different title pages are carefully studied and explained. According to López Piñero and López Terrada the reasons for the differences in the title pages can be explained by the Inquisition's censorship. Since the Inquisition had prohibited the work the title page could not refer to "Fuchs's Herbal in the Romance language" and thus, Birckmann had to change the title page in order to circulate and distribute the translation.[24] Where editions of the Spanish translation of Fuchs are to be found today, such as the one at the University of Valencia they bare the marks of extensive censorship and expurgation.[25] It is most interesting to note that Lastanosa owned a work which had been censored by the Inquisition.

Lastanosa and His Interests in Botany, Agriculture and Gardening.

In reseaching the life and works of Lastanosa it is important to frame his achievements within the context the context of the history of scientific activity in Spain during the Sixteenth and Seventeenth centuries. In addition, we must also remember that Lastanosa had an extensive collection of botanical works in his library. However, we must not forget that botany as a seperate independent discipline did not exist as such but rather was considered to be part of natural history or materia medica. For a long time botany was also closelt associated with the study of agriculture and gardening. This is very much the clase when reviewing the works found in Lastanosa's library. As was the case with many collectors in Europe, Lastanosa was very interested in botany and in obtaining information about exotic plants from and the New World. This can be seen in his interest in the works of Nicolas Monardes and Francisco Hernández. But before speaking about these important figures in the history of Spanish science let us discuss another noted physician and humanist Andrés Laguna.

Lastanosa and Laguna

Andrés Laguna (c. 1510-1559), the famous medical humanist was born into a Jewish-Converso family, as was Miguel Servet and most probably Francisco Hernández. He was without a doubt one of the most important figures in the history of Spanish botany and materia medica. While he was born in the city of Segovia, he, like other academics of his time, spent most of his life on a continuous academic pilgrimage. Much like Servet, Laguna spent his life traveling, studying and living in various European cities such as Salamanca, Alcalá de Henares, Toledo, Paris, London, Cologne, Bologna, Pisa, Rome, Venice, Padua and many others. During his life he published more than thirty works dealing with various themes such as history, medicine, philosophy, literature and politics. In addition he wrote many commentaries on Greek and Latin texts. His most important medical works deal with the writings of Galen from a humanist perspective. However, his most famous work, which is even cited in Cervantes's *Don Quixote*, is his Spanish translation and commentary on the Materia medica of Dioscorides published in 1555. Even though this work is a rich source for the study of materia medica, we must recall that Laguna's work was based on the translation and edition of Dioscorides published by Mattioli in 1554. In addition, Laguna included many illustrations from the Mattioli in his own edition. Furthermore, we must observe that much of the information on New World plants found in Laguna is incorrect. Laguna's work is also significant in the area of the history of gardening in Spain, especially in terms of royal botanical gardens. In his work, Laguna advised King Phillip II to establish royal botanical gardens based on Italian models. This advice lead to the establishment of royal gardens such as those of Aranjuez.

Lastanosa, Gregorio de los Ríos, Gabriel Alonso de Herrera and Spanish Gardening

Lastanosa's interest in gardening can be clearly seen in the botanical and horticultural works, which he had in his personal library. This is especially seen in the works of Gregorio de los Ríos and Gabriel Alonso de Herrera (c. 1480-1539). Lastanosa had a unique interest in agriculture and this is testified the work of Gregorio de los Rios which he owned. Gregorio de los Ríos was a cleric gardener and chaplain to the Casa de Campo in Madrid. He is perhaps best remembered as one of the authors of one of the first modern works on gardening. This work is called *Agricultura de la jardinería* ('The Agriculture of Gardening'). This work was published by Pedro Madrigal in 1592. De los Rios spent is life working the gardens in what is today the Instituto Nacional de Investigaciones de Técnica Agaria (The Spanish National Institute of Agronomy). Among other tasks he also concentrated on the cultivation of aromatic and medicinal plants and herbs for the Royal Pharmacy.

Lastanosa's library also contained the work of another well-known agricultural writer. Here we are referring to Gabriel Alonso de Herrera. Alonso de Herrera was a priest horticulturalist, who was born between 1470 and 1480. He was also the brother of Hernando Alonso de Herrera (c. 1460-c. 1529), the noted Spanish humanist and professor of Rhetoric and Grammar at the Universities of Salamanca and Alcalá de Henares and author of *Obra de agricultura* (1513).[26] Alonso de Herrera was also the personal chaplain of Card. Jiménez de Cisneros who commissioned his work on agriculture. Gabriel Alonso de Herrera spent most of his life working in Granada where he was in charge of the farmlands of the Marques de Mondejón. We also know that he traveled widely through Spain, France and Italy studying the agricultural practices of those countries. The work of Gabriel Alonso de Herrera is based on Greek, Latin and Arabic sources, as well as the author's personal experiences. Within the history of Spanish gardening he is considered to be a Renaissance follower of the Medieval Arabic tradition of agriculture in Andalusia. His work deals with many agricultural themes and topics: soils, vineyards, trees, crops, agricultural animals and the agricultural calendar. His book was very popular. Many editions of his work were published in French, Latin and Italian. In addition, we must recall the fact that this work was published in 1513 and as such is one of the first works of its kind dealing with agriculture during the Renaissance and one of the first to be written in Spanish. However, we must point out that his work was not an original work but rather a compilation based on many sources. But nevertheless, this work by Alonso de Herrera was a rich source of information and advise for his fellow countrymen.

Apart for the above-mentioned works on agriculture Lastanosa also owned other works on this topic. Among these we can cite the works of Antonio Micaldo, *Secretorum agi* (Luteial 1560), Antonio Benuto, *De agricultura, Agricultura practica* (Barcelona 1626) and the work of F. Mugel Agustin. *Secretos de la agricultura* (Periñon 1626).

Lastanosa and Monardes

One of the most important botanical works found in the library of Lastanosa is that of Nicolas Monardes. Lastanosa owned Monardes's *De las cosas que se traen de las Indias...* (Sevilla 1569).[27] As we know, Monardes, the noted physician and botanist from Seville was born in 1493 and died in 1588. He is considered to be on of the founding fathers of Spanish botany and materia medica. Monardes is also remembered for his accomplishments in the areas of natural history, medicine and metallurgy.[28] Like his contemporary Francisco Hernández he studied at the University of Alcalá de Henares. His early works concerned roses, citrus, snow and iron. However, there is no doubt that his most important work was his *Historia medicinal de las cosas que se traen de nuestras Indias* published in 1574. This work is made up of three major sections. The first part was published in 1565 and the second in 1571. As we will soon see, the note physician from Seville placed a great deal of importance on his wonderful place of birth. Seville was a major port and site of embarkation for ships traveling to the New World and it was in that city that Monardes cultivated some of the first plants brought from the Americas in his famous garden. We know that other colleagues who also had gardens and museums in Seville frequently visited his garden. Among these we must mention Gonzalo de Argote de Molina and Simon Tovar.[29]

In addition, we must highlight the fact that Monardes was a central figure in the dissemination of information on New World plants and materia medica. Many European botanists and physicians used his works concerning both medicinal plants and foodstuffs from the Americas. Here we can cite information on maize, peanuts sarsaparilla and coca. Some of his botanical descriptions became classics such as those on the Balsams of Peru and Tolu. It is clear that Monardes's work holds a central place in the history of botany during the Sixteenth century. His works as well as translations of his works such as those by Clusius were European best sellers. They were translated into more than six languages and appeared in more forty editions.[30] Before leaving Monardes, we would like to make use of this opportunity to mention the recent interesting work of Professor Daniela Bleichmar concerning his work. In her essay, "Books, Bodies and Fields. Sixteenth Century Transatlantic Encounters with New World Materia Medica,"[31] Dr. Bleichmar approaches the study of Monardes and his work from the perspective of the importance of market relations and the network of commerce between Europe and the New World. She underscores the importance of the study of natural history, and specifically material medica through the lens of mercantile relations. According to Bleichmar "Colonial botany" was practiced not only in the Americas but also in the courts, palaces, gardens, battlefields, physician offices and pharmacies through out the world.

Plants and Flowers in Lastanosa's Garden

And so, what was Lastanosa's garden like? What plants and flowers were cultivated in his gardens? We know that during the Sixteenth century ornamental gardens were first established in Spain. Frequently, in addition to medicinal plants these gardens contained collections of exotic plants as well as native plants. Using documents concerning Lastanosa's garden we know that there were many flowering plants. Among others, his magnificent garden contained hyacinthines, narcissi, tulips, roses, carnations, lilies and jasmines.[32] Lilies and roses held special places of honor in Sixteenth and Seventeenth century gardens. This was especially true in the gardens of Phillip II.

Maize, Chile, Narcissi and Tulips: Plants from the New and Old World

We know that exotic plants from Asia, as well as those from the Americas were highly regard and were very much the fashion in European gardens since the Sixteenth century. The plants were appreciated for both there medicinal as well as their ornamental values.

Maize

Some of the earliest descriptions of maize (*Zea mays* L.) are found in the writings of Columbus, Alvarez Chanca, Hernando Colón, Fernández de Oviedo, Cabeza de Vaca, López de Gómara, Ruelle, Bock, Monardes, Mattioli, and of course Fuchs.[33] Fuchs referred to maize using the name *Turcicum frumentum* ('Turkish grain').[34] However, one of the first and most detailed descriptions of maize, along with an excellent illustration of the plants is found in the work of the German botanist H. Bock in 1552. However, the clarification of the confusion as to the exact origin, American rather than Asian, had to wait until the diffusion of the work of Fernández de Oviedo and Monardes, especially through the translations and editions of Clusius. Only then were the American origins of maize were firmly established. It was not until 1576 that botanists from Northern Europe became aware that maize was not from Turkey. This first came to light in the works of M. Lobel (O'Lobel or Lobelius). During this time Francisco Hernandez was in New Spain (Mexico) carrying out his research on Mexican medicinal plants. It is interesting to review his early description of maize. In his description, Hernández used the Nahuatl word *Tlaolli* in referring to maize. Here we will cite the Accademia dei Lincei edition of Hernandez, the edition which Lastanosa owned in reviewing the description of maize:

> "…alij, tlaolli, quod nostri frumentum Indicum, Haitini uero maizium appellant."[35]

> ["…that which we call Indian wheat and the people of Haiti call maize.][36]

Further on in his description, Hernández addes:

> "…a quibus recentiorum frumentum Turchichum, ab alij consultius Indicum appellantum.."[37]

> ["…some recent authors have called it Turkish wheat and others more correctly Indian wheat…"][38]

Chile

References to Chile (Chilli) or "Ají" abound in the early literature of the descriptions of New World plants. Some of the first references are found in the writings of Columbus, Alvarez Chanca, Hernando Colón, Fernández de Oviedo, Peter Martyr, Cienza, López de Gómara, Monardes, Hernández, and Fuchs.[39] Chile (*Capsicum annum* L.), that signature Mexican plant is one of the five plants from the New World that Fuchs described. However, as it was the case with maize, Fuchs was unaware of its American origins. In his description he referred to the plant as *Silquastrum.*[40] As we know this term comes from Pliny.[41] In his description Fuchs refers to three varieties: *Siliquastro maius et minus*; *Siliquastro terium* and *Siliquastro quartum.* Concerning Chile, Fuchs writes:

> "It is found in all parts of Germany. It is grown most frequently in flower pots and it is a recent arrival to Germany".[42]

Flowers in Lastanosa's Garden

Among the many plants found in Lastanosa's garden we find Narcissi. We know that thes flowers were highly regarded since the time of Antiquity for their many uses, especially thier oramental uses.[43] We will recall that the name Narcisus relates back to the ancient Greek legend whose most frequently cited version is to be found in Ovid's *Metamorphoses*. We will also remember that in the study of plant symbolism it is frequently associated as a symbol of vanity and egoism.

According to the Valencian art historian López Terrada and specialist in Spanish floral paintings, the genus Narcissus has more than forty species.[44] López Terrada has noted that these plants commonly appear in Spanish flower paintings and still lifes. The most frequent representations are those of the species known as *Narcissus tazetta* L., *Narcissus poeticus* L. and *Narcissus x ordorus L.*[45] She also observes that these flowers are frequently found in the paintings of the Spanish artist Juan de Arellano.[4647]

Fuchs and Exotic Plants

Fuchs's work is full of information on exotic plants from both the New and Old World. His important manuscript, known as The Vienna Codex, is an significant source for exotic plants which had been introduced into Germany from the Middle East and the Americas. Among those from the Middle East we find a great deal of information on tulips, that highly esteemed flower. We will recall that the first important illustrations of these plants are found in the work of Mattioli (1565) and in Fuchs. In referring to the tulip Mattioli used the term *Caryophyllus indicus maior.*[48] Fuchs's illustrations of the tulip might actually be earlier than those found in the Italian's work.[49] What are known today as tulips *Tulipa gesneriana* L. are represented by four varieties in the work of Fuchs: stá representada con cuatro variedades en la obra de Fuchs: *Bulbus sylvestris maior, seu anomone pratensis Theophrasti punicea*; *Bulbus sylvestris maior, seu anomone pratensis Theophrasti lutea*; *Bulbus sylvestris maior, seu anomone pratensis Theophrasti pallida*; and *Bulbus sylvestris maior, seu anomone pratensis Theophrasti candida.*

There is quite a bit of information on the early illustration of the tulip. According to López Terrada, the first description and illustration of the plant appears in the *Hortus Germaniae Liber* (1561) by the Swiss botanist Gesner. We also know that around 1581, M. Lobel published a description of the tulip with many descriptions. However, we will recall that the first illustration of this plant appeared in the work of R. Dodoens in 1569. Nevertheless, we must observe that the first and perhaps most imortant scholar of tulips was Clusius. In his work *Rariorum plantarum Historia* published in Antwerp in 1601. In this work, Clusius offers us many details and precise scientific information on th etulip and its varieties. In addition, Clusius also published other botanical illustrations of the tulip between 1583 and 1601.

Tulips in Spain

Tulips were cultivated in many places ain Spain but no where more than in the Royal Gardens such as those of Aranjuez. Again we must underscore the interest which Phillip II had in the establishment of these gardens, as well the work of Flemish gardeners in their maintenance. It seems more than sixty varieties of tulips were introduced and cultivated in these gardens. In our brief commentary on the tulip in Spain we must not forget to mention the importance of the *Codex Pomar*, a work that Lopez Piñero has studied in detail. This document is significant not only because it contains illustrations on flora and fauna of both the Old and New World but also because it offers information on the work of Francisco Hernández. Here we will only which to recall that in the text we find illustrations of five varieties of tulips: Tulipán rubea, Tulipán punicea, Tulipán alba, Tulipán alba perfoliata and the Tulipán punicea perfoliata.[50]

Tulips in Spanish Painting

Ornamental plants of Asian origin are frequently represented in the flower paintings of the Seventeenth century.[5152] One of the most common plants represented is of course the tulip. As López Terrada has indicated with great detail and care, there are many examples of the representations of tulips in both Dutch and Spanish paintings. As we know the tulip was of central importance in Baroque flower vase painting. This importance is seen in their symbolic function. Here we are referring to the allusion of the transitory and ephemeral nature of human life.

Lastanosa and his Tulips

Documentation concerning Lastanosa's life and works indicate that Françios Filhol (1583-1648), the noted French collector and close friend of Lastanosa sent him tulips, or rather seeds and bulbs from France:

> "Los tulipanes que Francia cria, /I tu curiosidad prodiga embia, / desarrollan aqui vistosamente/tanto esplendor luciente,/i tanta variedad, que no ai colores/que puedan dibuxar sus resplandores".[53]

We don't know of any studies on the history of tulips in Spain.[54] Nor have we found any information of the history of the tulip in Spain or France in any of the works, which we have consulted.[55] However, the recent work of Miguel López on the friendship between Lastanosa and François Filhol do offer some interesting insights into the history of the plant especially as far as France and Spain are concerned.[56] What we do know was that there was something known as "Tulipmania"[57] which spread all through France at the beginning of the Seventeenth century, not only in Paris but also in the north and south of France.[58] According to Dash:

> "…around 1610 there was a craze for flowers in Paris, and fashionable nobles began competing with each other to present the ladies of the court with the rarest and most spectacular specimens they could find".[59]

Frequently, these flowers were so valued that they were used in the place of jewels.[60] It is said that at the wedding of Louis XIII in 1615 it was very much in fashion for aristocratic ladies to wear tulips in their *décolleté* as jewels, which more valued than diamonds.[61] The fad and obsession for tulips in France could also be noted in the popularity of flower paintings such as those of Nicolas Robert (1614-1685)[62], as well as in the vast numbers of catalogues and treatises on tulips.[63] According to some Frenchmen, the marvelous character and nature of the tulip could only be described by using the language of alchemy.[64]

Spanish Baroque Paintings and Flowers

At the beginning of this essay we underscored the important links between art and science, especially botany in the life of Lastanosa. We emphasized the importance of reviewing the work of Ferrari in this respect. Now we will turn more in detail to flower painting. Since Lastanosa had many books on botany, agriculture and gardening, many with extensive plant illustrations such as those of Ferrari and Hernandez to mention just two. We can well imagine that he loved the artistic representation of flowers and flower paintings in particular. We know that he had one such painting, "Some Flowers by Camillo."[65] In order to draw an end to our essay, we would like to review the works of some Baroque painters and their flower paintings. As we know, in these paintings we find many representations of exotic plants from Asia and the Americas.[66] As has been observed we can clearly see the presence of plants from the New World in Baroque painting.[67] In these painting we can note both the ornamental value of these plants as well depictions of their food value. Take for example the painting by the Spanish artist Murillo (1619-1682) that is entitled "The Kitchen of The Angeles."[68] In this painting we find images of common foods, which were to be found in the Spanish kitchen of the Seventeenth century. In this painting we can clearly identify peppers, tomatoes and other plants recently introduced to Spanish cuisine.[69] Francisco Hernández provides us with descriptions of many of the ornamental and food plants that are found in Baroque paintings. Many of these plants would also appear in the works of Spanish painters during the Baroque period. Among these ornamental plants we find, sunflowers, tagetes, nasturtiums, nards and passionflowers.[70]

A close look at the works of some of these flower painting, such as those of Juan de Arellano (1614-1676) will provide us with some clear idea of the contents of these paintings.[71] Arellano was one of the most important flower painters of the Spanish Baroque. His works show the clear marks of Italian and Flemish influence. One of his works called "Flower Garland with the Allegory of Vanity" (1646) is a good example of his work and one, which Lastanosa might have been familiar with. This is one of Arellano's first important paintings of his mature period. It seems that this painting was carried out with the assistance of Francisco Camilio. Might this be the same Camilo or "Camilio" whose work was found in the Lastanosa collection? M. J. Lopez Terrada has provided us with careful identifications of some of the flowers found in this beautiful example of Spanish flower painting.[72] Using her identification we will cite a few examples of some of the plants found in the painting that might also be found in Lastanosa'a garden. Among these we find: lilies, tulips and narcissi.[73]

Conclusion

In conclusion we can only say that there is much more work to be done in this area of study. Surely there are other flower painters from Aragon whose works need to be examined in this context. Surely there are other historians of art and historians of botany who no more about Lastanosa's art and botanical works than we do. It is our opinion that in order to have a better understanding of Lastanosa'a life, work and collections they must be approached by an interdisciplinary group of scholars working together. What we are calling for is the creation of a new Lastanosa Circle who will work together and exchange information about one of Huesca's most famous native sons.

Notes

1. Juan Francisco Andrés de Uztarroz, *Soneto del Doctor Juan Francisco Andres ablando con el Padre Gracian*, Hispanic Society of America, Ms. B-2424.
2. Baltasar Gracián, *The Art of Worldy Wisdom.*, (Springfield, Illinois: Charles C. Thomas, 1954). English translation by Martin Fischer.

3. See John F Healy, "Introduction", in: *Pliny The Elder. Natural History. A Selection,* (London: Penguin, 1991), Ix-xli and R. French, "The Natural History of Pliny" in: *Ancient Natural History. History of Nature*, (London: Routledge, 1994), 196-255. Lastanosa had varoius editions and translation of Pliny in his library.
4. Javier Ordóñez, Víctor Navarro y José María Sánchez Ron, *Historia de la ciencia*, (Madrid, Espasa-Calpe, 2004), 178.
5. Ordóñez, Navarro & Sánchez, *Historia*, 269-270.
6. See Paula Findlen, *Possessing Nature: Museums, Collecting, and Scientific Culture in Early Modern Italy*, (Berkeley, University of California Press, 1994).
7. See Findlen, *Possessing Nature*, 4-5 and her recent edition of essays and studies on Kircher, Paula FINDLEN, ed., *Athanasius Kircher: The Last Man Who Knew Everything*, (New York, Routledge, 2004). See also the collection of essays by Oliver Impey and Arthur MacGregor, eds., *The Origins of Museums: The Cabinet of Curiosities in Sixteenth- and Seventeenth-Century Europe*, (Oxford, Clarendon Press, 1985). There is very little on the life and work of the Danish physician Ole Worm. The best is perhaps the work of H. D. Schepelern, *Museum Wormianum. Dets Forudsaetninger og Tilblivelse* (København, 1971). At the present time we are working on a study of the influence of the work of Francisco Hernández on the work of Ole Worm.
8. Richard Lightbown, "Some Notes on Spanish Baroque Collectors," in: Oliver Impey and Arthur MacGregor, eds., *The Origins of Museums: The Cabinet of Curiosities in Sixteenth- and Seventeenth-Century Europe*, (Oxford, Clarendon Press, 1985), 136-46. Besides Lastanosa, Lightbown discusses the collections of Diego Felipe María de Guzmán, Geronimo Fures y Muñoz, Juan de Espina, Francisco Ximénez de Urrea, Ribaldo de Valencia among others. It should be said that the majority of the collections discussed are art or painting collections.
9. Lightbown, "Some Notes", 139. Even a quick look at the *Descripción en verso* (1647), will indicate that Lastanosa had an extensive museum: "…admirasse en su estancia/tanta diversidadi extravagancia/de cotas. De lorigas i arneses, /de espada, de rodelas…tienen una alacena por clausura/diversos esqueletos/de aves, de pescados i animales, /crystal de roca, candidos corales…". We are also told that he had various fossils and even the "horn of a unicorn."
10. On Ferrari see, Lucía Tongiorgi Tomasi, " 'L'arte ingenua e ingengnosa di cultivari i fiori.' Note su *Flora Overo Cultura di Fiori* de Giovan Battista Ferrari", in: Giovan Battista Ferrari, *Flora ovvero Cultura di fiori,* (Roma: Leo S. Olschki, 2001), ix-xxv; A. Campitelli, "Gli 'Horti di Flora'. I Giardini di Roma al tempi di Giovani Battista Ferrari", in: Ferrari, *Flora,* xxvii-xliii; D. Freedberg, *The Eyes of the Lynx*, (Chicago, University of Chicago Press, 2002), 38-39 and 46-57; C. Serafini, *Arte, scienza diletto nello Roma de Urbano VIII: il trattato De Florum Cultura di Giovanni Battista Ferrari S. I. (1583-1655). Una fonte per il giardino italiano*, (Ph.D. Diss. Universita degli Studi de Roma La Sapenza, 2004-2005).
11. Freeberg, *The eyes*, 15-38 and 53-64.
12. Giovanni Battista Ferrari, *De Florum Cultura Libri IV*, (Roma, Stephanus Paulinus, 1633); Giovanni Battista Ferrari, *Flora ovvero Cultura di fiori*, (Roma, P. A. Faciotti, 1638); Giovanni Battista Ferrari, *Hesperides, sive De Malorum Aureorum Cultura et Usu Libri Quattour*, (Roma, Herman Scheus, 1646).

13. Carolus Clusius, *Rariorum aliquot Stripium…*, (Antwerp, Plantin, 1583); Rembert Dodoens, *Stirpium historiae penptades sex. Sive libri XXX*, (Antwerp, Plantin, 1583).
14. *Narración de lo que le pasó a Don Vincencio Lastanosa a 15 de octubre del año 1662 con un religioso docto y grave*. Hispanic Society of America (HSA), Ms. B-2424, 52r-79vº. (56v; 76v.-77v).
15. N. Van Hout, "A Second Portrait in Rubens's *Four Philosophers*", *The Burlington Magazine,* 142 (November 2000): 694-697.
16. See Anne Goldgar, *Tulipmania. Money, Honor and Knowledge in the Dutch Golden Age,* (Chicago, University of Chicago Press, 2007), 34.
17. From the works of Lipsius, Lastanosa had in his library: Justii Lipsii, *Saturalium Romanorum libri duo,* (Antuerpiae, 1584) and Justii Lipsii, *De amphiteatro*, (Antuerpiae, 1585).
18. Ferrari, *De Flora*, 377.
19. Ferrari, *De Flora*, 371-393.
20. Freedberg, *The eyes*, 38-40.
21. G. B. Marini Bettolo, "L'Iconografia del Tesoro", in: *Una Guida alla Lectura de Tesoro Messicano, Reum medicarum Novae Hispaniae*, (Roma, Instituto Poligrafico e Zecca dello Stato, Liberia dello Stato, 1992), 39-40.
22. Unfortunately, the epigraph "Fathers of German botany" has been greatly disseminated and has created a mythology in the history of science. It seems that this term has its origin in the work of C. Sprengel (*Geshichte der botanik,* 1817) and later spread by E. H. F. Meyer (*Geshichte der botanik*, 4 vols., 1854-1857) and J. Sachs, (*Geshichte der Botanik,* 1860). On this topic see, F. J. Anderson, *The Illustrated History of the Herbals*, (New York, Columbia, 1977), 51. But in our opinion López Piñero y Pardo have addressed this issue more clearly: "Mucho más conocida es la referencia que menciona los desenfoques causados por los acercamientos nacionalistas. El tópico de considerar 'padres de la botánica' a los alemanes Leonhart Fuchs, Otto Brunfels y Hyeronimus Boch ha sido desmentida por las investigaciones acerca de su directa dependencia del mundo rencentista italiano", see José María López Piñero and José Pardo Tomás, *La influencia de Francisco Hernández (1515-1587) en la constitución de la botánica y la materia médica modernas,* (Valencia, Instituto de Estudios Documentales y sobre la Historia de la Ciencia, Universitat de Valencia, 1996), 31.
23. On the life and works of Fuchs, see E. Stüber, *Leonhart Fuchs, Heben und Werk* (Munchen, 1928); T. A. Spague and E. Nelma, "The Herbal of Leonhart Fuchs", *Botanical Journal Of the Linnean Society of London,* 48 (1931): 545-642; José María López Piñero and María Luz López Terrada, *Las traducciones por Juan de Jarava de Leonhardt Fuchs y la terminología botánica castellana del siglo XVI*, (Valencia, Instituto de Estudios Documentales y sobre la Historia de la Ciencia, Universitat de Valencia, 1994), 20-30; José María López Piñero and María Luz López Terrada, La *influencia española en la introducción en Europa de las plantas Americanas (1493-1623)*, (Valencia, Instituto de Estudios Documentales e históricos sobre la Ciencia, Universitat de Valencia, 1997), 35-39. Among more recent works see A. Pavord, "The Irascible Fuchs", in: *The Naming*

of Names, The Search for Order in The World of Plants, (New York, Bloomsbury, 2005), 175-204.

24. López Piñero and López Terrada, *Las traducciones*, 34-35.

25. See for example the edition found in the Biblioteca Historicomédica de Valencia, López Piñero and López Terrada, *Las traducciones*, 43.

26. Gabriel Alonso de Herrera, *Obra de Agricultura copilada de diversos auctores*, (Alcalá de Henares, Arnao Guillén de Bocar, 1513).

27. Nicolás Monardes, *Dos libros. El uno trata de todas las cosas que se traen de las Indias occidentalis, que sirven al uso de medicina...* (Sevilla, Hernando Díaz, 1569). It was No. 709 in the catalogue of Lastanosa's library.

28. Francisco Rodríguez Marín, *La verdadera biografía del Dr. Nicolas Monardes*, (Sevilla, Padilla, 1925); J. Lasso de la Vega y Cortezo, *Biografía y estudio crítico de las obras del médico Nicolas Monardes*, (Sevilla, Padilla, 1988); José María López Piñero, "El autor y su obra", in: Nicolás Monardes, *La Historia medicinal de las cosas...(1565-1574)*, (Madrid, Ministerio de Salud y Consumo, 1989), 13-22.

29. On Tovar, see José María López Piñero, *Diccionario Histórico de la Ciencia Moderna en España,* (Barcelona, Península, 1983), I: 371-373. Tovar had an important correspondence with Arías Montano (a close friend of Francisco Hernández), Bernardus Paludanus from Leiden and more importantly, with Clusius.

30. On Clusius and his relations with Monardes and the dissemination of information on plants from the Americas see, López Piñero and López Terrada, *La influencia*, 65-68.

31. Daniela Bleichmar, "Books, Bodies, and Fields. Sixteenth-Century Transatlantic Encounters with New World Materia Medica", in: Londa Schiebinger and Claudia Swan, eds., *Colonial Botany. Science, Commerce, and Politics in the Early Modern World*, (University of Pennsylvania Press, 2004), 83-99.

32. Uztarroz, *Soneto*.

33. López Piñero and López Terrada, *Las traducciones*, 144-148; 301.

34. Frederick Meyer, Emily Trueblood, and John Heller, eds., *The Great Herbal of Leonhart Fuchs: De Historia Stirpium Commentarii Insignes, 1542: Vol 1 & 2.* (Stanford, Stanford University Press, 1999). 623-626.

35. Raquel Álvarez Peláez, *De Materia medica novae Hispaniae Libri Quatro. Cuatro libros sobre la materia medica de Nueva España. El manuscrito de Recchi*, (Aranjuez, Doce Calles, 2000), 602-608. See also the Mexican edition of the work of Francisco Ximénez, *Cuatro Libros de la naturaleza,* (México, 1888), 194-200 and the Spanish translation in Francisco Hernández, *Historia de las Plantas de Nueva España,* in: *Obras Completas,* (México, UNAM, 1959), II: Libro VI, Cap. XLIV, 288-292 and finally our English translation: Simon Varey, "First Texts", "Chili", "Corn", in: Simon Varey and Rafael Chabrán, *The Mexican Treasury*, (Stanford, Stanford University Press, 2000), 107-113; 111-113.

36. Álvarez, *De materia medica*, Libro VII, Cap. XL. Our translation.

37. Ibidem.

38. Hernández, *Historia*, 288-292. See also our English translation: Varey and Chabrán, *The Mexican Treasury*, 111-114: "We shall say then that *tlaolli*, which

we may call maize (the more familiar and more common word for it) for some people today a Turkish grain, for others more properly and Indian one…"
39. López Piñero and López Terrada, *Las traducciones*, 172-174; 259-260.
40. Fuchs, *Historia*, 731.
41. Plinio, *Historia naturalis*, Book 19, Chap. 187. On this point it is also interesting to note Francisco Hernández 's translation and commentary of Pliny, in Francisco Hernández, *Historia natural de Cayo Plinio Secundo,* (Madrid, Visor, 1998), 800 and 823.
42. See, Fuchs, *Historia,* 731.
43. M. J. López Terrada, *Tradición y cambio en la pintura Valenciana de flores (1600-1850),* (Valencia, Universitat de Valencia, 2001), 64.
44. López Terrada, *Tradición*, 65.
45. Ibid.
46. Ibid.
47. For this study we have used the edition of this work as found in Frederick Meyer, "The Vienna Codex: Leonhart Fuchs's Unpublished Herb Book", in: Meyer, Trueblood and Heller, *The Great Herbal*, 147-194.
48. Meyer, Trueblood and Heller, *The Great Herbal*, 168.
49. On the tulip in Fuchs's work see, Meyer, Trueblood and Heller, *The Great Herbal*, 168-169.
50. López Terrada, *Tradición*, 80-81.
51. López Terrada, *Tradición*, 76.
52. On this topic see, Goldgar, *Tulipmania*, 90-92.
53. Francisco Andrés de Uztarroz, *Descripcion de las antiguedades i jardines de Don Vincencio Juan de Lastanosa…*, (Zaragoza, Diego Dormer, 1647), 15.
54. The only information on tulips in Spain is found in López Terrada, *Tradición*, 77-79.
55. A. Pavord, *The Tulip. The Story That Has Made Men Mad*, (New York, Bloomsbury, 1999); M. Dash, *Tulipomania. The Story of the World's Most Coveted Flower and the Extraordinary Passions It Aroused*, (New York: Three Rivers Press, 1999).
56. On Lastanosa's circle, see Miguel López Pérez's contribution to this volume.
57. Pavord, *The Tulip*, 73-89.
58. Dash, *Tulipomania*, 67.
59. Dash, *Tulipomania*, 65.
60. Pavord, "The Irascible Fuchs", 73.
61. Dash, *Tulipomania*, 66.
62. Pavord, "The Irascible Fuchs", 74.
63. Pavord, "The Irascible Fuchs", 75.
64. Ibid.
65. *Narración*, 76-77.
66. López Terrada, *Tradición*, 81.
67. López Terrada, *Tradición*, 80.
68. María Luz López Terrada, "Hernández and Spanish Painting in the Seventeenth Century", in: Simon VAREY, ed., *Searching for the Secrets of*

Nature, (Stanford, Stanford University Press, 2000), 153-154 and López Terrada, *Tradición*, 81, note 174.

69. López Terrada, *Tradición*, 81. See also our essay on the history of Spanish food, Rafael Chabrán, "Medieval Spain", in: Melitta Weiss Anderson, ed., *Regional Cuisines of Medieval Europe. A Book of Essays*, (New York, Routledge, 2002), 143-146.

70. López Terrada, *Tradición*, 81, note 174 and López Terrada, "Hernández".

71. On Arellano see López Terrada, *Tradición*, 90-93.

72. López Terrada, *Tradición*, 92-93.

73. López Terrada, *Tradición*, 93.

CHAPTER TWELVE

VINCENCIO JUAN DE LASTANOSA, TULIPS, AND SEVENTEENTH-CENTURY COLLECTING

ANNE GOLDGAR

Because of the fame, or rather notoriousness, of the Dutch tulipmania of the 1630s, one of the few things that a more general readership might know about early modern gardens is that they contained tulips. Even works about tulips in gardens in other countries – Spain, for example – tend to lapse into talk of tulipmania in the Netherlands.[1] However, there is much more to be said about tulips than simply to wonder at high prices paid for them. In my recent work I have tried to place the possession of tulips, even at lower social levels, into a wider culture of aesthetic appreciation, development of hierarchies of knowledge and expertise, and the building (and sometimes destruction) of social communities based around discussion and exchange of tulips and other plants.[2] As many other historians and historians of science have noted, the circulation of knowledge and of objects, including plants, in the late sixteenth and seventeenth centuries is part of a set of complex trends, ranging from a growing relationship among mercantile, aesthetic, and scientific interest in natural historical objects to the connection of knowledge to wealth and status.[3]

I would like to set Lastanosa in this context: the context of tulips as an object of collection, discussion, and display. Where Lastanosa is concerned, we are in a similar position to the general public: one of the few things we know about his garden is that it contained tulips. Much of what we know about Lastanosa's garden comes from the descriptions by the chronicler Juan Francisco Andrés de Uztarroz, published in verse in 1647, and recast in manuscript in prose in 1650.[4] Uztarroz was apparently much more interested in the wonders of the house, including its collection of antiquities, art, natural history objects, scientific instruments, maps, and exotica. When it came to the garden, especially in the longer account of 1650, Uztarroz treated it as something separate, an appendix to the lengthy description of the house and its contents. He discussed the statues in the garden, with their allegorical references to its contents – one statue, for example, was of Flora, and another of Vertumno, the spouse of Pomona[5] – and took considerable space discussing the shape and layout of the garden and its geometrically-arranged paths and beds. But he wrote only in the most general terms about the actual contents of the garden: vague phrases like *diversos arboles y plantas*, "various trees and plants," suggest that botanical collecting was not Uztarroz's main interest.[6] We get a somewhat more specific impression of the flowers in the garden from the verse description of 1647, which was dedicated to the French collector and cleric François Filhol, who had sent Lastanosa tulips for his garden. From this we hear in more in detail about the beauties of the tulips, narcissi, jasmine, and roses in the garden, as well as about other plants. But this praise is also in the most general terms, and certainly more eulogistic than purely descriptive.

Uztarroz was interested in form. But I would like to speak instead about content, and not so much physical content, but the social and cultural content of gardens, something of which tulips are emblematic in this period. In some ways what is most interesting about Lastanosa's garden is not the thing itself, but that Uztarroz and others decided to describe it and that people like François Filhol gave Lastanosa flowers to plant in it. Indeed, it is interesting that Lastanosa chose to have a garden at all.

In late sixteenth-century and seventeenth-century culture, gardens played a variety of social and cultural functions. Gardens were sites of discussion, exchange, and the display of status. Knowledge about plants, including and sometimes especially tulips, was for a certain community intrinsically fascinating, and for that community, and other circles overlapping with it, such knowledge could carry with it a degree of status. I have studied these sorts of communities particularly in the Netherlands of the period, but we can see similar trends in other parts of Europe, such as the Holy Roman Empire, France, Italy, and also in Spain. Lastanosa, then, was part of a much broader culture of interest in tulips and other flowers and plants in the period. Not all collectors and plant enthusiasts were the same, however. In the Netherlands, for example, at just the period Lastanosa began collecting, there was a growing community of expertise and specific knowledge about plants among burghers and master craftsmen.[7] Although conditions in the Netherlands at the time of tulipmania were unusual, in all the places mentioned the real experts on gardens were generally professional gardeners, and when tulip bulbs were stolen, which happened frequently in the late sixteenth century, it was usually the gardeners who were suspected, as they knew the value of the bulbs. We need to consider why Lastanosa – or others – had tulips in their gardens at all. In doing so, we also need to consider the types of communities that valued tulips in the late sixteenth and seventeenth centuries and why.

It is of course no accident that Lastanosa's house contained a collection of *naturalia* and *artificialia* and also looked out onto an extensive garden. Collecting and gardens went together. The early fashion for collecting at the Habsburg courts – under Rudolf II, for example – began a trend for combining interior and exterior reflections on the possible interactions of art and nature.[8] While collections frequently contained objects that joined together the artistry of nature and the artistry of man – nautilus shells etched with pictures of other sea creatures, for example, or coconut shells carved with biblical or mythological scenes[9] – the garden put the arrangement of nature and its propagation in the hands of humanity. It was frequently the possibility of nurturing and even creating new types of flowers that excited interest among early enthusiasts for exotic plants. The Flemish plant collector Jehan Boisot bragged in 1590 that his tulips that year had shown a good deal of novelty, something for which he was naturally to be thought responsible.[10] As Erik de Jong and others have pointed out, the formality of gardens like Lastanosa's can be attributed to a desire to control a nature that was in other respects virtually uncontrollable.[11] Gardens, like collections of *naturalia* and exotic objects, also gave collectors a chance to contemplate the wonders of God and nature. We therefore find collections frequently paired with gardens; botanical gardens, such as the one at the University of Leiden, often came with an adjoining collection, and similarly many private collectors, from Rudolf II to small fry like the apothecaries Paul and Jacques Contant in Poitiers, combined preserved *naturalia* and works of human craft with a living garden.[12] Charles de Croy, the duke of Aerschot, transported Jacques Plateau and his family from Tournai to Beaumont to take charge of both his cabinet and his "foreign plants, from which I take great pleasure and delectation, admiring the works of nature so much even though I have so many other occupations of importance...." Aerschot expressed a desire in 1604 for "anything strange, be it plants, animals, or other rarities."[13]

Aerschot's word "strange" – *estrange* – is important here. Because of the greatly expanded botanical universe of the sixteenth and seventeenth centuries, with thousands more plants coming into Europe than had ever been seen before, gardens, like collections of ethnographic objects, also gave collectors reason to think about the possibility of worlds beyond those described by the ancients. In one of the chamber towns of the Dutch East India Company, Enkhuizen, the first important Dutch collector, the doctor Barent ten Broecke or Bernardus Paludanus, built his collection around materials coming into the Netherlands from abroad, both botanical and otherwise. From around 1600, Paludanus was able to profit from his close connection with the explorer Jan Huygen van Linschoten, who seems to have provided most of the exotic materials in Paludanus' collection, items from India, China, and Japan.[14] Paludanus' collection, which was visited by many well-known scholars and aristocrats despite its relatively obscure situation in Enkhuizen,[15] was well-known for its garden, and when Leiden was looking for the first director of its new botanical garden, its curators tried (unsuccessfully) to hire Paludanus precisely because he combined botanical knowledge with a collection, which the curators wanted him to bring with him to Leiden.[16]

The importation of new plants that helped make Paludanus' collection interesting was of course even more important in Spain. As many historians of Spanish science have emphasized, the Spanish colonial enterprise had a distinctly scientific and utilitarian component, with royal expeditions, such as that of Francisco Hernández, directed specifically at searching out new plants for medical and commercial use.[17] From the Codex Pomar and the works of Spanish botanical writers such as Monardes we know the degree to which plants from the New World proved of interest to those at home, and the news of these new plants travelled north through the work of people like Carolus Clusius, whose *Rariorum aliquot stirpium per Hispanias observatarum historia* of 1576 informed the rest of Europe of the plants, local and exotic, in the gardens of Spanish botanists such as Simón de Tovar in Seville.[18] Clusius also publicized the work of botanists of the Indies, such as Acosta, Monardes, and Orta, in his *Exoticorum libri decem* of 1605.[19]

Many of these figures were more interested in useful plants, such as *materia medica*; but it has been noted that Clusius and many others had a large place in their conception of the botanical world for exotic flowers, such as those praised by Uztarroz in Lastanosa's garden. Tulips, irises, narcissi, crocuses, hyacinths, crown imperials, fritillaries, and many other flowers, frequently larger and more dramatic in appearance than the old medieval European flowers, became the particular focus of many gardens.

They also fuelled the new genre of floral still life, which grew up around the same time in the Netherlands and in Spain in the late sixteenth and early seventeenth centuries. Although fewer still lifes in Spain featured flowers, those that did, such as works by Juan van der Hamen y León or Antonio Ponce, portrayed brightly-colored tulips.[20] Flemish floral still lifes, along with most other genres of Flemish painting, were also sought after in Spain, and these even more frequently contained tulips.[21] Although still life as a genre was not popular among theorists of art in Spain, it became popular with collectors in the period Lastanosa was collecting – he himself owned several – in part because the style was considered modern.[22] And as floral paintings by artists such as Van der Hamen y León were in demand, we might suggest that real tulips were also of interest in Spain.[23] Certainly the flowers were in Spain from at least 1567, when one of the gardeners brought by Philip II from Flanders to design the gardens at Aranjuez, Jehan Holbecq, imported from the Netherlands many sorts of exotic bulbous plants. These included the tulip, described by him in a list of goods as "a flower that comes from Constantinople and that one can find in the garden of Cardinal Granvelle" (whose gardener was Holbecq's brother François).[24] Tulips were at this time of the utmost rarity even in Flanders.

Although most of the ornamental flowers of interest to the Spanish were American, the tulip, like many other plants entering Europe in the mid-sixteenth century, came not from the New World, but from Turkey and originally Asia Minor. It proved of particular interest both to botanical writers and to more general collectors and floral enthusiasts because of its aesthetic qualities, especially its appeal to the early modern appreciation for variety and diversity. Unlike most other flowers, Clusius and other botanical investigators found, a plain-colored tulip grown from seed might, after seven to ten years, suddenly change into something more splendid, with speckles, stripes, flames, and other variegated patterns that delighted the early modern sensibility. Botanists and collectors were uncertain of the reason for this – it was generally believed that the grower could affect the change, although in fact it appears that changes happened through mutation, accidental crossbreeding, and disease – but the possibility of change as well as the exotic origins of flowers made them ideal for collectors. An enthusiastic breeder could, apparently, stamp his own identity on a tulip by turning it into something more beautiful and more exciting than previously, and something moreover that was different from what other people possessed.

This had a number of results for those who wished to see themselves as collectors. The exotic origins of tulips – as well as of other rare flowers

– were of considerable interest to early collectors, who tended to repeat stories about the early encounters with tulips of the imperial envoy to the sultan, Ogier Ghislain de Busbecq, in the 1550s. Especially in the early years, the rarity of tulips was also highly exciting for people interested in flowers. Some of the literary productions in France, Flanders, and the Netherlands about tulips portrayed other, more traditional flowers such as roses and daisies as reacting petulantly to being superseded by another flower rarer and more special than they were. Botanists such as Clusius were besieged with requests for tulips, especially rare ones. We also can see that the possibility of having what someone else did not was highly appealing to collectors aspiring to collections that were not encyclopaedic, but rather focused on the unusual and the strange. Some possessors of tulips took care to make sure their particular breeds did not fall into the hands of others. One Dutch monthly newspaper reported in 1623 that a certain collector in Amsterdam had all twelve examples of the rare and valuable Semper August bulb and would not consider selling;[25] and similarly at mid-century the French courtier Nicolas Valnay lamented in print the stubbornness of another collector who would not share the riches of his collection of rare breeds.[26]

If beauty, variety, exoticism, and rarity were all attractions to collectors, another factor was just as important, and that was a feature not entirely of the tulip, but of the collector as well. Tulips appealed to collectors because of the knowledge that was required to own a really good collection. One had to know something about how to cultivate tulips and how to promote them to change (although, as previously stated, this was not in fact possible), and one also had to know what the best varieties were and where to acquire them. The very fact that some varieties were considered better than others demonstrates to us that a body of knowledge was developing about tulips and a set of standards being applied to them, in the same way that connoisseurs of art were beginning in the seventeenth century to develop more specific critical faculties to judge what was good and bad in painting. Tulips were like most collectibles in the period, because to be able to speak with authority about them, to appreciate their beauties and to know about their relative importance and qualities, would give a collector status within a certain type of community – a community that valued that type of knowledge. The history of collecting in recent years has focused on the way that the possession of curiosities as well as the viewing of them provided opportunities to display status, and it is evident that tulips fell within a similar framework.[27]

Some of those who owned tulips were part of such a group. As we saw earlier, the combination of collections of curiosities and rarities frequently went hand in hand with a live garden as well as the dried plants to be found in *herbaria* indoors. Given that Lastanosa had a fair-sized collection of paintings, including paintings attributed to Netherlandish masters,[28] it is worth noting that in the Netherlands many of the same people who bought tulips were also collectors of paintings. Evidence from inventories and auction records allows us to be quite precise about this in the case of Amsterdam, where it is tempting to speculate that the kind of community that developed around the relatively frequent sales of art at auction led to an exchange of knowledge and a communication of the relative equation of values between rare flowers and rare paintings.[29] Certainly at an auction of the estate of an Amsterdam florist in 1626, in which the flowers sold were chiefly anemones, we find bidding by large numbers of art collectors, including those who we know were buying expensive tulips a few years later.[30] In Paris, similarly, it is easy to identify major collectors, such as Gaston d'Orléans, the brother of Louis XIII, who combined all these kinds of collecting. Orléans, who is claimed by the (now agreed to be falsified) account of Lastanosa's house and gardens, "Las tres cosas mas singulares que tiene la Casa de Lastanosa en este año de 1639,"[31] as a contact of Lastanosa's, combined all these kinds of collecting. Orléans was considered by Antoine Schnapper to be emblematic of what he was trying to say about French collecting in his classic studies *Le Géant, la licorne, la tulipe* and *Curieux du Grand Siècle*.[32] He began a flower garden at his home in Blois in the 1630s, which later developed into an important botanical garden; he also commissioned a famous series of vellum portraits of his flowers by Nicolas Robert, as well as possessing a large collection of other artworks, medals, and other curiosities. Sixteenth- and seventeenth-century Flemish collectors, such as Rubens and Nicolaas Rockox, also brought together large collections of paintings and important gardens containing tulips.[33]

In content, then, we can make some comparisons between Lastanosa and others who lived around his time. Like them – and like others in Spain, especially those at court – Lastanosa combined a collection with a garden, and like some of them, he also combined a collection of paintings with a garden containing tulips and other flowers grown for beauty rather than utility. Although we cannot be sure of how Lastanosa saw or used his garden, we can extrapolate to some degree from other cases to suggest that it might have prompted contemplation, among other things on the relationship between art and nature and on the wonders of God. Miguel López Pérez has recently argued that one purpose of the garden might have been to combat melancholy, not simply of Lastanosa personally, but of the pervasive melancholy anatomized by Thomas Burton at the beginning of the seventeenth century.[34] We know, especially from the work of Mark Morford,[35] that gardens were favored by the neostoics, and therefore were an appropriate pursuit for the patron of Gracián. Lipsius, for example, wrote with approval in *De Constantia* of the function of gardens as places of retirement and contemplation, and although he frowned slightly on the excessive attachment of some of his neostoic friends to particular gardens or plants, he made clear his adherence to the value of gardens for allowing one to contemplate the wonders of the world in microcosm.[36]

On the other hand, as Lipsius made clear by discussing gardens in a dialogue with several characters, gardens were places of sociability as well as solitude. The sociability associated with gardens and tulips is strongly linked to the creation of both knowledge and status. The comparison with collections, or the inclusion of tulips within collections, makes this clear. Just as collections gave one the opportunity both to learn, to make intellectual connections, and also to acquire and demonstrate status through one's knowledge, so did gardens function in similar ways. Nor did this necessarily have to happen at a high social level. Where tulips in Holland were concerned, such discussion, transfer of knowledge and acquisition of status could occur fairly low down in society, at the level of what I call "tulip experts," who spent considerable time in the 1630s talking to each other in gardens. Such experts were something like art dealers, a profession a few of them actually practiced, and their discussions appear to have touched on the properties of valuable tulips, the technicalities of how such tulips should be bred, who owned which items, what kind of prices were reasonable for particular flowers, and other such topics.[37] This is of course a demonstration of some of the features of the commodified nature of the sociability associated with tulips at this period. But, besides stressing that such sociability almost always had a commodified aspect, I would like to emphasize that this kind of discussion is a reflection of the sociability also to be found in gardens at a higher social level. We find it in the humanistic gardens of noble and burgher correspondents of Clusius in Brussels and Tournai in the later sixteenth century, who made it clear in their letters the degree to which they and other floral enthusiasts were trading botanical information, much of which they passed on to the Clusius and some of which was included in his botanical writings.[38] We find it as well in the gardens of courtly Parisian tulip collectors such as Nicolas Valnay or the Sieur de la Chesnée Monstereul, where, these writers tell us, we would find enthusiasts in the springtime flying to and fro to compare each other's flowers and to discuss the desirable qualities and the novel aspects of the latest blooms.[39]

Besides exchanging knowledge, such discussion once again helped to identify which collectors had status, status gained through the possession of information, the possession of taste, or simply the possession of rare tulips. As in collections and museums in the early modern period, simply being admitted to the discussion was a way to gain status, again, of course, within a community that specifically valued that kind of status. An aristocrat whose rationale for a person's acceptability was focused on the quality of his ancestry would not be particularly impressed by the possession of rarities or ability to converse with the knowledgeable. This was a significant point in the United Provinces, where to a great extent those in the aristocracy were not very interested in rare flowers. It was a different matter in the southern Netherlands, however, a fact that will prove of relevance to the case of Spain. A standard pattern for the beginnings of a great courtly collection – and most of the great collections in Spain in this period were courtly – was diplomatic service abroad, either in the Netherlands or in Italy. Spain's aristocracy, or an important part of it, was indeed interested in those who could discuss art and curiosities, so that a collection and the sociability that accompanied it could prove of importance to status within a more traditional hierarchy.[40]

We know that Lastanosa valued sociability; the existence of his circle makes that clear. It is even evident that his sociability could play a role within the context of his collecting. This was clearly true when we consider his numismatic collecting, where it is evident that discussion with members of his circle such as the count of Guimerá in Zaragoza, Uztarroz, Urrea, and others, who all published on numismatics, played an important role in the development of his own views.[41] What is less obvious is to what degree this circle was likely to discuss the plants in Lastanosa's garden. As we have noted, Uztarroz, for one, seems when writing descriptions of Lastanosa's collection not to have been very interested in the botanical aspects of his garden. This is strikingly different from the kind of account one finds when reading the letters and other writings of collectors who prided themselves on a specific familiarity with particular breeds and names of tulips and other flowers. On the other hand, the accounts by Uztarroz alert us to another aspect of the sociability available to Lastanosa. We know from his verse dedication to François Filhol, as well as from other sources, that Lastanosa was in correspondence with Filhol and received gifts of tulips from him.[42] Filhol, a canon at the cathedral in Toulouse, was linked into the collecting and gardening circles of France; he was listed by Pierre Borel as one of around fifteen important collectors in Toulouse in the period, and he was sufficiently connected with Gaston d'Orléans that he presented him with a catalogue of his varied collection (which included skeletons and *écorchés*), a manuscript that survives in the Bibliothèque Nationale de France.[43] Filhol had various connections with Aragon; he corresponded with friends of Lastanosa, including both Uztarroz and the antiquary Francisco Jiménez de Urrea, and Uztarroz in fact translated Filhol's description of his own collection into Spanish in 1644.[44] Filhol was not the only connection Lastanosa had with French collecting. Among his other correspondents were Jean-Baptiste Dru, a merchant in Lyon and known as herborist of the king of France, who published catalogues of his own plants in 1649 and 1653 containing close to 550 varieties of tulip.[45] Many of the tulips in Dru's catalogue, moreover, were originally developed by the most important professional florists in Paris, Jean Robin and Pierre Morin, both of whom came from families who had long been in the flower business.[46] Morin was yet another of Lastanosa's correspondents, and in 1655 Lastanosa received from Morin the printed catalogue of the Parisian gardener's own tulips.[47] Morin, we are told by Uztarroz, was one of the most important experts on tulips in Europe and was, moreover, a well-known collector of other items; his cabinet in the midst of his oval garden in Saint-Germain, containing notably shells and butterflies, reflected the aesthetics of the tulips in the garden outside. If Lastanosa was not in contact with many in the world of gardening and floral expertise, he was in touch with some of the right people.

We can extend the idea of "tulip experts" – those who know the most about tulips, their cultivation, and their aesthetic and monetary value – to talk about nodes within a web of contacts and discussion. Just as a tulip expert standing in a garden in Haarlem might command particular authority to those standing near him as he pronounced on the value of particular tulips or the way particular colorations might be achieved, so were there experts within the European community of collecting and floriculture who, through their own work with plants and their contacts with others, might distil the knowledge of scattered floral enthusiasts and perhaps redistribute that information in speech, in correspondence, or in print. Someone like Carolus Clusius in the sixteenth century, or perhaps Pierre Morin in the seventeenth, would count as such as a node. Clusius, who is said – probably incorrectly – to have brought the tulip to the Netherlands, was subject to a flood of correspondence about tulips, containing questions about how they should be grown, who had what new varieties, and giving other pieces of important news to an intellectual community without any central means of communication. He was also asked by many of his correspondents for bulbs, seeds, and cuttings, which he distributed across Europe. As we have noted, Clusius was in contact with botanists in Spain in the late sixteenth and early seventeenth centuries, following on from his travels in Spain in mid-century, and the information people like Simón de Tovar gave him ended up in his writings.[48]

It is not easy to place Lastanosa within this world of floral expertise. Although we can cite a few names of experts with whom Lastanosa apparently exchanged correspondence, we cannot be sure of the quality of those contacts. That Morin sent Lastanosa a catalogue of his flowers, for example, is not necessarily significant. The book was widely distributed and was indeed a sale catalogue, dedicated to enlightening his public what kind of bulbs were available in his garden in Paris. Although Lastanosa had close connections in Aragon and in Madrid, evident from his dealings over other kinds of collecting, we cannot assume from the statements in Uztarroz' accounts or the "Narración de lo que le pasó a Don Vincencio Lastanosa a 15 de Octubre del Año 1662" that he had significant contacts outside Spain. The assumption in many books that he had visited and corresponded with Gaston d'Orléans is based on the now-discredited document "Las tres cosas mas singulares que tiene la Casa de Lastanosa en este año de 1639," and there is no further evidence that such contacts existed. Except for Morin in Paris, the correspondents Uztarroz mentions were all just on the other side of the Pyrenees or in Italy. It is in fact interesting, given the wide association of the Netherlands with the tulip after the much-publicized tulipmania of 1636-7, that Uztarroz identified tulips specifically with France ("los Tulipanes que la Francia cria") when celebrating their existence in Lastanosa's garden.[49] The overall impression is that Lastanosa and his friends did not have a wide net of contacts concerning the gardens in Huesca. Indeed, we must be careful not to make too much of contacts with other international scholars either. The fact that Lastanosa wrote seven letters to Athanasius Kircher, for example, must be seen in the context of Kircher's typical style of correspondence, in which Kircher either asked for information or was asked for it, but there was rarely a real exchange. In this case, Lastanosa's letters were dominated by his desire to obtain Kircher's books, with little other content.[50] It is difficult to describe this as Lastanosa's existing within a web of exchange of information, or to see Lastanosa's floral world as being an international one.

Although one can easily envision the connections between Lastanosa's collection of curiosities and his garden of tulips, it is not immediately evident that Lastanosa approached his garden as a means of contemplating nature, and in particular as a means of learning about rare plants. The botanical gardens of Spain might best be represented by the correspondents of Clusius at the end of the sixteenth century. Simón de Tovar, for example, was so concerned with the minute changes in his garden and the significance these might have had to his fellow-botanists that he published yearly catalogues that he distributed throughout Europe. Although we are necessarily subject to the incompleteness of our sources, what we read in Uztarroz's and Lastanosa's own accounts suggests that there was none of this kind of attention to detail in Huesca; and although Miguel Morán and Fernando Checa, who in 1985 discussed the relation of cabinets of curiosities and collections, concede that Lastanosa's garden can be called a "collection," they nevertheless suggest that its purpose was more the curiosity and the exploration of diversity than the progress of knowledge.[51] Certainly if we look at Lastanosa's library, we see few books about the cultivation of plants. Although he owned editions of Dioscorides, Acosta, Monardes, and Fuchs, he possessed none of the northern European herbals published after the middle of the sixteenth century, thereby denying himself much of the recent and constantly-changing knowledge available there about the subject of botany. Although we learn from the 1662 account of his house that he had acquired some of the modern books about flowers, such as La Chesnée Monstereul's *Le floriste françois* and Giovanni Baptista Ferrari's *Flora, seu florum cultura*, it is notable that these books were written largely for a more courtly market, with Monstereul most anxious about display and exclusivity, and Ferrari indulging in frequent classical allusion in text and engravings, without devoting much space to the technical aspects of floriculture. That the author of the 1662 description (possibly Lastanosa) was not well-versed in the botanical literature is clear from his mistaking the *Hortus Floridus* of Crispijn van de Passe, a book consisting almost entirely of engravings and intended for a moneyed and leisure market, with works by the botanists Rembertus Dodonaeus and Carolus Clusius.[52]

The form of the garden as described by Uztarroz and others also does not give the impression of a place devoted to the study of botany. In Spain of this period we can make a strong distinction between botanical gardens and classical gardens; the latter denoted the fashion from the mid-sixteenth century of making gardens extensions of classical collections indoors. Such gardens displayed sculptures, medals, reliefs, and inscriptions, giving an *al fresco* expression to very similar curiosities within the attached houses. These kinds of classical gardens were first seen in the villas of Italian aristocrats, but the fashion in fact made its way to Spain chiefly through the Italianate Flemish gardens designed by artists such as Hans Vredeman de Vries and viewed by the future Philip II during his sojourns in Flanders in the middle of the sixteenth century.[53] We might argue that Lastanosa's flowers were merely decorative, although their rarity was an attraction, perhaps making the decoration even more decorative. Such flowers would have been a foil to the display of erudition and taste evident in the sculpture also to be found in the garden. It is probable that at the very least Lastanosa saw his garden as a necessary accompaniment to his collection of rarities, his paintings, his library, and his convening of a cultural and literary circle in Huesca. The sociability and status accruing through this would have stood him in good stead in his position as an aristocrat making a life for himself in the provinces.

A comparison that immediately springs to mind is Peiresc.[54] The neostoic scholar Nicolas-Claude Fabri de Peiresc, a lawyer at the parlement of Aix, came from a provincial noble family. If being a center of a correspondence network makes one a node of expertise, Peiresc was certainly that; his exchanges with a large number of other European correspondents about everything from antiquities to astronomy made him one of the best-connected provincial scholars of the era. On the other hand, he never published any of his works, although he did write essays on a variety of subjects. Among the things that fascinated Peiresc was flowers. He followed the gossip of the French community of tulip-enthusiasts with interest, and his letters to his brother in the 1620s depict an active life of collecting tulips and anemones, as well as a familiarity with and dealings with flower dealers in Provence and Pierre Morin and Jean Robin in Paris ("Maurin will accommodate you more than Robin," he advised his brother in 1625).[55] Peiresc was making a life of learning in the provinces – a point one could make as well about someone like the apothecary Paludanus, whose collection made the unlikely location of Enkhuizen a tourist attraction for the learned and the aristocracy of northern Europe at the end of the sixteenth century. Lastanosa was also trying to make such a life, but Peiresc in fact was much more successful in his cultivation of a wide circle of learning in the first part of the seventeenth century. However, it may also have been more his aim to do so than it was Lastanosa's.

When thinking about the role of tulips in gardens and in collections, we need to think about this last point: what the aim of the collector or owner actually was. Peiresc comes closer to Lastanosa in social standing than someone like Paludanus, who at the end of the day was a small-town apothecary, actually the professional identity of many early modern collectors. Elizabeth Hyde wrote in a perceptive article in 2002 about flowers, taste, and class in seventeenth-century France that it is not easy to classify the status and roles of those who had tulips in their gardens. She notes the difficulty in using a distinction between bourgeois and aristocrat in looking at flower-collectors; some were merchants and craftsmen, some were doctors and apothecaries, and others were bureaucrats, aristocrats, and, as we have seen in the case of Gaston d'Orléans, even princes of the blood.[56] Hyde suggests that an interest in flowers helped to break down social barriers by redefining social identities and roles. Florists (as collectors were called) were required to be knowledgeable about flowers, to appreciate their aesthetic qualities, and to follow the fashion in plants. This insight provides us with food for thought as we consider the social mixing and the way that expertise brought status within the world of tulip-lovers. The value of a humanistic education among an important section of the European aristocracy is a sign of the increasing respect in the period for knowledge and for those who produced it. The concentration, for example, of a figure like Philips Marnix van St. Aldegonde, the second-in-command of William of Orange during the Dutch Revolt, on tulips prompts a number of reflections. One such reflection might be on the role of neostoicism: that someone like Marnix wanted to seek solace in his garden in the midst of conflict. But another is that an aristocrat thought it perfectly natural, even desirable, to be in contact with scholars of no social standing. In the same way, the fact that noblemen like Lastanosa and Peiresc were in correspondence, and a valued correspondence, with apothecaries about plants tells us something about the way certain hierarchies were being reshaped, at least in some spheres, in the early seventeenth century.[57]

Such an analysis must be placed against what has already been said here about the emphasis in the seventeenth century on the interplay of knowledge and display. Lastanosa, as a provincial nobleman who fashioned himself as a learned man, will have wanted, at least on some level, to cultivate his image on the European stage of letters. The flowers he planted in his garden, but even more the fact that they came as gifts from learned foreign collectors, facts noted in all the descriptions of his garden, will have helped to shape that image. Having a garden, having tulips in it, having a collection, talking to scholars, writing letters, receiving gifts, and even receiving letters: these all helped to turn the possession of tulips into a new and different kind of status for a provincial nobleman like Lastanosa.

The comparison with French and Dutch collectors, however, should leave us wondering about whether Lastanosa was indeed completely equivalent to other small provincial collectors across Europe. Although we can stress the neostoic character of Lastanosa's apparently melancholic collecting, not to mention his patronage of Gracián, we should also note the contradictory pull of the court in a country where collecting was a fashion, indeed under Philip IV almost an obsession, led by the monarchy. The fact that one of the apparently false descriptions of Lastanosa's house and gardens, the "Tres cosas más singulares," probably written by a descendent, chooses to exaggerate the importance of the gardens not only by claiming contacts with Gaston d'Orléans, but also by pretending that Philip IV and other important nobles had visited the collection,[58] perhaps gives some indication of the importance to the Lastanosas of courtly favor.[59] Spain had already seen in Philip II a monarch deeply interested in collecting curiosities, art, scientific instruments, and plants, which he valued both for their ornamental and their scientific properties. Philip II created gardens, such as those at Aranjuez, that set a fashion complementing collections of other items. Although Philip III had been less interested in collecting than his father, his favorite Lerma was a major collector, and as Lerma was responsible for bringing the high aristocracy back to court, he also helped to encourage a fashion for collecting that took hold among the court nobility. Under Olivares, major collections grew up at the top of society, not least among Olivares' own family and clients.[60] Philip IV has been called a "mega-collector"; according to Jonathan Brown his collection of pictures was indubitably the largest in seventeenth-century Europe.[61] Although it was certainly not necessary to be at court to be a collector – scholars have noted the existence of collectors from the mercantile and professional classes[62] – it is clear that what Jonathan Brown and J.H. Elliott have called a "veritable epidemic of collecting" in Lastanosa's period was set off by the crown.[63]

The scholarship on this fashion for collecting has concentrated on the collection of art, but although the truly great collectors such as Gaspar and Luis de Haro or the Marquis of Leganés concentrated on art more than the more universal collections of the sixteenth century,[64] noble and courtly collections continued to echo the interest in *Wunderkammer* objects we find in Lastanosa's. Philip II's taste for exotica, natural history objects, medals, and scientific instruments affected the taste of a later generation even as Philip IV's passion for paintings increased the interest in this form of collecting.[65] Although the lack of focus to date by the scholarly literature on non-artistic aspects of the noble Spanish collections makes it difficult for us to compare the contents of Lastanosa's collection of *Wunderkammer* specimens with the collections of his contemporaries, we can see from his paintings that in at least some areas he was indeed collecting at least in part in imitation of courtly taste. Just as Philip IV and the major aristocratic art collectors focused on northern and Italian art, including a mania for Rubens,[66] we hear from the accounts by Uztarroz and others that Lastanosa's collection of paintings and prints contained Italian works by Titian, Caravaggio, Raphael and Tintoretto, and Flemish or northern works by such artists as Bartholomeus Spranger, Lucas van Leyden, and Dürer, not to mention two copies of paintings by Rubens.[67] Although we cannot trust the attributions in this list – the names of Lucas van Leyden and Dürer, for example, were frequently assigned to any northern artist[68] – we can see that to some degree Lastanosa was influenced by the taste of the court in Madrid and the great aristocratic collections. In other ways, it must be admitted, he followed his own taste; Lastanosa concentrated much more heavily on religious works than did the great collectors, who gravitated strongly toward secular painting, and his collection paid more attention to Spanish artists than his potential courtly models, who as we have seen demonstrated a strong preference for the foreign.[69] Although we therefore cannot make an entirely clear connection between Lastanosa's taste and that of the court, it is evident, especially from the works mentioned in the short list provided in the 1662 description of the house in Huesca, that he was at least aware of and to some degree echoed courtly taste. This taste was also partly that of an earlier age – Morán and Checa surmise that he had not made a full transition to a modern style of collecting[70] – but it was nevertheless the taste of the monarchy.

If we look at the gardens in Huesca we come to a similar conclusion. As we have already noted, gardens in Spain in the seventeenth century were influenced strongly by the Italianate style established in the peninsula by Philip II in imitation of those created by garden designers in Flanders. An antiquarian sculpture garden such as that described by Uztarroz sounds very similar to other aristocratic gardens modelled on sites such as the Pardo, Aranjuez, and Buen Retiro. Like these, Lastanosa's garden featured classical references, statuary, long avenues lined with topiary, labyrinths, and floral borders.[71] Although we do not have much information about the floral content of such classical gardens, we can surmise that tulips formed at least some part. The presence of tulips in most Spanish floral still life of the period – a style of painting apparently considered a sign of modern taste[72] – is only one of many indications that tulips were valued in the period. The planting of tulips in royal gardens from the age of Philip II indicates that courtly gardening, although focusing on exotic plants from theAmericas, would make possession of the tulip yet another way of imitating royal taste.[73] The emphasis on tulips in the descriptions of Lastanosa's gardens indicates his own pride in them, and together with his apparent emulation of other aspects of royal collecting, we might speculate that the tulips and the garden themselves represented yet another manifestation of an eye on the fashions of the court.

The paintings in his collection, and the statues and the tulips in garden, can all help us to understand the motives for Lastanosa's collecting. Although the imitation of the collections and gardens of Castile was not slavish, it has enough in common for us to suggest that his interest was not simply in a refuge, a *museo discreto* that provided an escape from a world of cares. His world and terms of reference were admittedly chiefly local, but – and the very fact that so many descriptions of his collection were produced underlines this – he was also keen for status. In one way, as we have seen, he was anxious to be a figure on the stage of the European learned world, as the prominence of names such as Filhol, Morin, and Dru in these descriptions indicates. The relative paucity of contacts outside his region, however, suggests that these aspirations were not particularly successful, as does his library's betrayal of weaknesses in his scholarship, in particular a lack of fluency in Latin.

But we can also see from his reasonably faithful emulation of courtly trends in collecting, especially in his imitation of the taste of Philip II, that if he was not actually visited by Philip IV and major courtiers, as the "Tres cosas más singulares" claims, such visits were in fact desired. In many European countries, collecting, as a preserve largely of the higher social levels, could be a means of social advancement. This was even more the case in a country like Spain, where collecting was such a passion for Philip IV and his nobility.[74] If the Aragonese Lastanosa was not himself located at the center of things, his collection might be viewed as partly a means to connect him to the court and its concerns. His dedication to Philip IV and to Juan José of Austria of his own few works and those by Gracián published by him are further indications of such an appeal.[75] In his later years, Lastanosa was able to forge links with Juan José, who arrived in Zaragoza in 1669 as Viceroy of Aragon, and in the wake of this he gained minor political office, taking charge of the archive of the kingdom of Aragon, becoming first magistrate of the local council, and, certainly through Juan José's influence, in 1677 becoming gentleman of the household of Charles II.[76] Juan José was known for his cultivation, and it is conceivable that shared interests in collecting and scholarship provided opportunities, however minor, for Lastanosa once the royal family was within easy reach of Huesca.

Whether we focus on tulips, art, scientific instruments, or wonders of natural history, we can find a range of reasons why Lastanosa might have had formed a collection. The accounts of his garden and of his cabinet by Uztarroz and others stress the wonders of his possessions as well as his pride in his own ancestry, and we must not dismiss Lastanosa's own fascination with knowledge and his sense of his family's worth as important reasons to collect. Lastanosa was, however, well-connected within Aragon, and we can deduce from the descriptions of his collection and other aspects of his life that he aspired to greater things. One of these, relatively unsuccessful, was to play a larger figure on the European learned stage. Another, it seems, was to appeal to courtly sensibilities, both through a particular style of collecting and through dedications of books. Whether the goal was political advancement or simply to follow the fashion we cannot be certain, but when Lastanosa's tastes coincided with those of important grandees who paid him attention, political office, even if relatively unimportant, seems to have followed. Juan José, after all, owned four paintings attributed to Rubens.[77] Lastanosa, on the other hand, had to content himself with two copies after Rubens. Perhaps it was not much, but for a minor provincial nobleman it was certainly better than nothing.

Notes

1. For example, María Luz López Terrada, "Las plantas ornamentales," in Carmen Añón and José Luis Sancho, eds., *Jardín y Naturaleza en el reinado de Felipe II* (n.p.: Sociedad Estatal para la Conmemoración de los Centenarios de Felipe II y Carlos V, 1998), 307-327.
2. Anne Goldgar, *Tulipmania: Money, Honor, and Knowledge in the Dutch Golden Age* (Chicago: University of Chicago Press, 2007). On social networks and tulips, see also Anne Goldgar, "Poelenburch's Garden: Art, Flowers, Networks, and Knowledge in Seventeenth-Century Holland," in Amy Golahny, Mia Mochizuki, and Lisa Vergara, eds., *In His Milieu: Essays on Netherlandish Art in Memory of John Michael Montias* (Amsterdam: University of Amsterdam Press, 2006), 193-191. Some of the arguments about tulips and collecting made later in the article can be found in expanded form in *Tulipmania*.
3. For a useful introduction to these debates, see Pamela H. Smith and Paula Findlen, "Commerce and the Representation of Nature in Art and Science," in Smith and Findlen, eds., *Merchants and Marvels: Commerce, Science, and Art in Early Modern Europe* (New York and London: Routledge, 2002), 1-25.
4. Juan Francisco Andrés de Uztarroz, *Descripcion de las Antiguedades I Jardines de don Vincencio Iuan de Lastanosa* (Zaragoza: Diego Dormer, 1647); Hispanic

Society of America, New York (henceforth HSA), Ms. B-2424, ff. 24r-51v, "Descripción del Palacio y los Jardines de Vincencio Juan de Lastanosa," 1650.
5. HSA, Ms. B-2424, Uztarroz 1650 account, f. 48r.
6. Ibid.
7. See Goldgar, *Tulipmania*, 131-191; and for humanistic collectors in the late sixteenth century, 20-61.
8. This is much discussed in the literature on collecting. See, for example, Erik de Jong, "Nature and Art: The Leiden Hortus as 'Musaeum,'" in Tjon Sie Fat and Erik de Jong, eds., *The Authentic Garden* (n.p.: Clusius Foundation, 1991). 37-70. The relationship between gardens and collections in Spain is discussed extensively in Miguel Morán and Fernando Checa, *El Coleccionismo en España* (Madrid: Ediciones Catédro, 1985), 139-204.
9. On this see especially Martin Kemp, "'Wrought by no Artist's Hand': The Natural, the Artificial, the Exotic, and the Scientific in Some Artifacts from the Renaissance," in Claire Farago, ed., *Reframing the Renaissance: Visual Culture in Europe and Latin America 1450-1650* (New Haven and London: Yale University Press, 1995) 177-196.
10. University of Leiden Library (henceforth UBL) Ms. BPL 1886, Jehan Boisot to Clusius, Brussels, May 17, 1590.
11. Erik de Jong, *Natuur en Kunst: Nederlandse Tuin- en Landschapsarchitectuur 1650-1740* (Bussum: Thoth, 1993), 35-41.
12. On Leiden, see De Jong, "Nature and Art"; for a description of the collection of the Contants, see Paul Contant, *Les Divers Exercices de Iacques et Paul Contant Pere et Fils Maistres Apoticaires de la ville de Poictiers* (Poitiers: Julian Thoreau and La Veuve d'Antoine Mesnier, 1628).
13. UBL Ms. Vulc. 101, Charles de Croy, duke of Aerschot, to Carolus Clusius, Beaumont, Sept. 25, 1604.
14. H.D. Schepelern, "Natural Philosophers and Princely Collectors: Worm, Paludanus and the Gottorp and Copenhagen Collections," in Oliver Impey and Arthur MacGregor, eds., *The Origins of Museums: The Cabinet of Curiosities in Sixteenth- and Seventeenth-Century Europe* (Oxford: Clarendon Press, 1985), 120; Roelof van Gelder, "Paradijsvogels in Enkhuizen: De relatie tussen Van Linschoten en Bernardus Paludanus," in Roelof van Gelder, Jan Parmentier, and Vibeke Roeper, eds., *Souffrir pour Parvenir: De Wereld van Jan Huygen van Linschoten* (Haarlem: Arcadia, 1998), 30-50.
15. Koninklijke Bibliotheek, The Hague, Ms. 133 M 63, album amicorum of Bernardus Paludanus.
16. De Jong, "Nature and Art," 47.
17. For an overview of the role of the monarchy, Juan Pimentel, "The Iberian Vision: Science and Empire in the Framework of a Universal Monarchy, 1500-1800," *Osiris* 2nd ser. 15 (2000), 17-30; more specifically, see, among many others, Jorge Cañizares-Esguerra, "Iberian Colonial Science," *Isis* 96 no. 1 (2005) 64-70; Antonio Barrera, "Local Herbs, Global Medicines: Commerce, Knowledge, and Commodities in Spanish America," in Smith and Findlen, *Merchants and Marvels*, 163-181; Antonio Barrera-Osorio, *Experiencing Nature: The Spanish*

American Empire and the Early Scientific Revolution (Austin: University of Texas Press, 2006); Paula De Vos, "The Science of Spices: Empiricism and Economic Botany in the Early Spanish Empire," *Journal of World History* 17 no. 4 (2006) 399-418; Simon Varey, Rafael Chabrán, and Dora A. Weiner, *Searching for the Secrets of Nature: The Life and Works of Dr. Francisco Hernández* (Stanford: Stanford University Press, 2000); David Goodman, *Power and Penury: Technology and Science in Philip II's Spain* (Cambridge: Cambridge University Press, 1988), chap. 5.

18. José M. Lopez Piñero and José Pardo Tomás, "The Contribution of Hernández to European Botany and Materia Medica," in Varey et al., *Searching for the Secrets of Nature*, 127; Nicolas Monardes, *Dos libros, el uno que trata de todas las cosas que se traen de nuestras Indias Occidentales, que serven al uso de la medicina, y el otra que trata de la Piedra Bezaar, y la Yerva Escuerçonera* (Seville: Hernando Diaz, 1569); Carolus Clusius, *Rariorum aliquot stirpium per Hispaniam observatarum historia* (Antwerp: Plantin, 1576).

19. Carolus Clusius, *Exoticorum libri decem* (Leiden: Officina Plantiniana Raphelengii, 1605).

20. William B. Jordan, *Spanish Still Life in the Golden Age, 1600-1650* (ex. cat. Fort Worth, Kimbell Art Museum, 1985), for example, plate 21; Jonathan Brown, *Painting in Spain 1500-1700* (New Haven and London, 1998), 89, 98.

21. Peter Cherry, "Seventeenth-Century Spanish Taste," in Marcus B. Burke and Peter Cherry, *Collectors of Paintings in Madrid 1601-1755* (Los Angeles: Provenance Index of the Getty Information Institute, 1977), I, 5; William B. Jordan and Peter Cherry, *Spanish Still Life from Velázquez to Goya* (ex. cat. London, National Gallery, 1995), 25.

22. Sarah Schroth, "Early Collectors of Still-Life Painting in Castile," in Jordan, *Spanish Still Life in the Golden Age*, 30.

23. William B. Jordan, *Juan van der Hamen y León and the Court of Madrid* (New Haven and London: Yale University Press, 2005), 74; Schroth, "Early Collectors of Still-Life Painting in Castile," 28, 30.

24. Albert Postma, "Een Zuidnederlandse hovenier in Spanje: Over Filips II, Jehan Holbecq en een nieuwe tuinkunst," *Tuinkunst* I (1995), 17.

25. Nicolaes Wassenaer, *Historisch verhael alder ghedenk-weerdichste geschiedenissen, die hier en daer in Europa...voorgevallen syn* VII (June 1624), 111-112.

26. [Nicolas Valnay], *Connoissance et Culture Parfaite des Tulippes rares, Des Anemones extraordinaires, Des Oeillets fins, Et des belles Oreilles d'Ours panachées* (Paris: Laurent d'Houry, 1688)

27. A classic exposition of these ideas is Paula Findlen, *Possessing Nature: Museums, Collecting, and Scientific Culture in Early Modern Italy* (Berkeley, Los Angeles, and London: University of California Press, 1994).

28. Ricardo del Arco y Garay, *La Erudición Aragonesa en el Siglo XVII en Torno a Lastanosa* (Madrid: Góngora, 1934), 50. The paintings are detailed by Uztarroz in his description of 1650, Hispanic Society of America Ms. B-2424.

29. On the connections between tulips and the collection of art, see Goldgar,

Tulipmania, 62-130, and John Michael Montias, *Art at Auction in 17th-Century Amsterdam* (Amsterdam: Amsterdam University Press, 1999), 70-76.
30. Gemeentearchief Amsterdam, WK 5073/952, auction of the flowers of Pieter Pietersz Tuynman, September 25, 1626.
31. Biblioteca Nacional de Madrid, Caja 18,727. The argument for this document's falsification is made by Fermín Gil Encabo, "La ficción 'telamoniana' de Pellicer en torna a Lastanosa," in Christof Strosetzki, *Actas de V Congreso de la Asociación Internacional Siglo de Ora* (Frankfurt and Madrid: Vervuert/ Iberoamerica, 2001), 623-634, and Fermín Gil Encabo, "Lastanosa y Gracián: en torno a Lastanosa," in Aurora Egido Martínez, Fermín Gil Encabo, and José Enrique Laplana Gil, eds., *Actas del I Congreso Internacional "Baltasar Gracián: pensiamento y erudición* (Huesca and Zaragoza: IEA/IFC/DGA, 2003), I, 19-60, and is discussed by Carlos Garcés Manau, "Vincencio Juan de Lastanosa: Una Biographia," in *Vincencio Juan de Lastanosa (1607-1681): La Pasión de Saber* (ex. cat. Huesca, 2007), 30. There is no other evidence that Gaston d'Orléans had contact with Lastanosa.
32. Antoine Schnapper, *Le Géant, la licorne, la tulipe: Collections et collectionneurs dans la France du XVIIe siècle* (Paris: Flammarion, 1988), and Schnapper, *Curieux du Grand Siècle: Collections et collectionneurs dans la France du XVIIe siècle* (Paris: Flammarion, 1994). Gaston d'Orléans features prominently in both volumes.
33. On Rockox, see Ria Fabri, "De Tuin van Nicolaas Rockox: 'un beau petit jardin garny de toutes sortes de belles plantes,'" in *Antwerpen een originele stadstuin rijker: Thien oraignie boomen sommigen met appelen belaeden* (ex. cat. Antwerp, Rockoxhuis, 2002).
34. Miguel López Pérez, "Anatomía del Virtuoso: Coleccionismo y Melancolía en la figura de Vincencio Juan de Lastanosa," *Argensola* 115 (2005), 144-164.
35. Mark Morford, "The Stoic Garden," *Journal of Garden History* 7, no. 2 (1987) 151-175; Mark Morford, *Stoics and Neostoics: Rubens and the Circle of Lipsius* (Princeton: Princeton University Press, 1991).
36. Justus Lipsius, *Two Bookes of Constancie*, trans. John Stradling (London: Richard Johnes, 1595), II, 63-64.
37. On tulip experts, see Goldgar, *Tulipmania*, 180-193.
38. Ibid., 25-29, 51-61.
39. [Valnay], *Connoissance et Culture Parfaite*; Sieur de la Chesnée Monstereul, *Le Floriste François, Traittant de l'Origine des Tulipes* (Caen: Eleazar Mangeant, 1654), distinguishes between "vrais Floristes" and those who claim to love flowers without understanding anything about them (pp. 191-202). On sociability and hierarchy in French flower-collecting of this period, see Elizabeth Hyde, "Flowers of Distinction: Taste, Class, and Floriculture in Seventeenth-Century France," in Michel Conan, ed., *Bourgeois and Aristocratic Cultural Encounters in Garden Art, 1550-1850*, Dumbarton Oaks Colloquium on the History of Landscape Architecture 23 (Washington: Dumbarton Oaks, 2002), 77-100.
40. On aristocratic and courtly collecting in Spain, especially of Italian and Netherlandish art, see, among others, Miguel Morán and Fernando Checa, *El*

Coleccionismo en España (Madrid: Ediciones Catédro, 1985); Marcus B. Burke and Peter Cherry, *Collections of Paintings in Madrid 1601-1755* (Los Angeles: Provenance Index of the Getty Information Institute, 1997); Alexander Vergara, *Rubens and his Spanish Patrons* (Cambridge: Cambridge University Press, 1999); Mary Crawford Volk, "New Light on a Seventeenth-Century Collector: The Marquis of Leganés," *Art Bulletin* 62 no. 2 (June 1980) 256-268; Jonathan Brown and Richard L. Kagan, "The Duke of Alacalá: His Collection and its Evolution," *Art Bulletin* 69 no. 2 (June 1987) 231-255.

41. Pablo Cuevas Subías, "El Circulo Lastanosino," in *Vincencio Juan de Lastanosa (1607-1681): La Pasión de Saber*, 186-187; on Lastanosa's numismatic collecting, Almudena Domínguez Arranz, "Monedas, Medallas y Piedras Preciosas en el 'Museo Discreto' de Vincencio Juan de Lastanosa," in idem., 125-133. On members of his circle and the numismatic tradition in Aragon, see Morán and Checa, *Coleccionismo en España*, 142.

42. Uztarroz, *Descripcion de las Antiguedades I Jardines de Don Vincencio Iuan de Lastanosa*, pp. 1, 3, 18; HSA Ms. B-2424, ff. 52r-79v, "Narración de lo que le pasó a Don Vincencio Lastanosa a 15 de Octubre del A o 1662 con un Religioso Docto y Grave," f. 78r; HSA Ms. B-2424, f.23, sonnet by Uztarroz, also published in Del Arco, *La erudición aragonesa del siglo XVII en torno a Lastanosa*, p. 83-84.

43. Bibliothèque Nationale de France, Ms. fr. 390. On Filhol, see Maurice Caillet and Robert Mesuret, "François Filhol (1583 (?)-1648. Son oeuvre et son cabinet," *Mémoires de la Société Archéologique du Midi de la France* 29 (1963), 99-137, and A. Coster, "Antiquaires d'autrefois. A propos de quelques lettres inédites de François Filhol, hebdomadier de Saint-Etienne de Toulouse, au chroniqueur d'Aragon Don Francisco Ximénez de Urréa," *Revue des Pyrénées* 23 (1911) 435-471.

44. Juan Francisco Andrés de Uztarroz, *Diseño de la insigne y bibliotheca de Francisco Filhol* (Huesca, 1944).

45. We learn of the connection with Dru in HSA Ms. B-2424, ff. 52r-79v, "Narración de lo que le pasó a Don Vincencio Lastanosa," f. 78r, where he mentions five men from whom he received flowers. On Dru, see Schnapper, *Le Géant, la Licorne, la Tulipe*, 50.

46. On Pierre Morin, see, among others, Prudence Leith-Ross, "A Seventeenth-Century Paris Garden," *Garden History* 21 no. 2 (winter 1993) 150-157.

47. The catalogue is mentioned in HSA Ms. B-2424, ff. 52r-79v, "Narración de lo que le pasó a Don Vincencio Lastanosa," f. 78r; it is probably Pierre Morin, *Catalogues de Quelques Plantes à Fleurs, qui sont de present au Iardin de Pierre Morin le jeune, dit Troisiéme; Fleuriste* (Paris: François le Cointe, 1651).

48. For a modern edition of some of this correspondence, see Josep Lluís Barona and Xavier Gómez Font, eds., *La correspondencia de Carolus Clusius con los científicos españoles* (Valencia: Seminari d'estudis sobre la ciència, 1988). On these connections, see, for example, José M. López Piñero and José Pardo Tomás, "The Contribution of Hernández to European Botany and Materia Medica," in Varey et al., *Searching for the Secrets of Nature*, 127-8.

49. Uztarroz, *Descripción de las Antiguedades I Jardines de Don Vincenio Iuan de Lastanosa* (1647), 15.
50. On Kircher's style of correspondence, see Noel Malcolm, "Private and Public Knowledge: Kircher, Esotericism, and the Republic of Letters," in Paula Findlen, ed., *Athanasius Kircher: The Last Man Who Knew Everything* (New York: Routledge, 2004), 298. Malcolm reports that 57% of Kircher's 763 correspondents only wrote to him once. On Lastanosa's correspondence with Kircher, including an edition of the letters, see Carlos Garcés Manau, "Diez Cartas de Vincencio Juan de Lastanosa y Diego Vincencio Vidania a Athanasius Kircher, Conservadas en la Universidad Pontificia Gregoriana de Roma," *Argensola* 115 (2005) 187-199.
51. Ibid., 149, 197.
52. See the edition of the inventory of Lastanosa's library, Karl-Ludwig Selig, *The Library of Vincencio Juan de Lastanosa, Patron of Gracian* (Geneva: E. Droz, 1960), 26, 38, 50. Reference to books on flowers is made in HSA B-2424 ff. 52r-79v, "Narración de lo que le pasó a Don Vincencio Lastanosa a 15 de Octubre del Año 1662," f. 59r. The entry on the *Hortus Floridus* states: "Rhemberto Dodoneo y Carlos Clusio, Hortus floridus, Arnhemii, 1614." The books by Ferrari and Monstereul are Giovanni Baptista Ferrari, *De Florum Cultura libri IV* (Rome: Stephanus Paulinus, 1633), and the Sieur de La Chesnée Monstereul, *Le Floriste françois*, cited above.
53. Morán and Checa, *Coleccionismo en España*, chap. 9, especially p. 143; Vicente Lleó Cañal, "Los jardines de la nobleza," in Carmen Añón and José Luis Sancho, eds., *Jardín y Naturaleza en la reinado de Felipe II* (n.p.: Sociedad Estatal para la Conmemoración de los Centenarios de Felipe II y Carlos V, 1998), 223-241, esp. 229, 236; Marta Nieto Bedoya, "El espíritu y la esencia de la jardinería en los Países Bajos," in ibid., 77-99; John H. Harvey, "Spanish Gardens in their Historical Background," *Garden History* 3 no. 1 (1974), 7-14.
54. On Peiresc, his intellectual style, and neostoicism, see Peter N. Miller, *Peiresc's Europe: Learning and Virtue in the Seventeenth Century* (New Haven and London: Yale University Press, 2000).
55. Peiresc to Palamède Valavez, Aix, April 26, 1625, in Philippe Tamizey de Larroque, ed., *Lettres de Peiresc aux Frères Dupuy* [etc.] (Paris: Imprimerie Nationale, 1888-1898) VI, 158. For more such letters, see vol. VI *passim*.
56. Hyde, "Flowers of Distinction," 81-4.
57. On these topics see also Brian W. Ogilvie, *The Science of Describing: Natural History in Renaissance Europe* (Chicago: University of Chicago Press, 2006).
58. That these visits apparently did not really take place is stressed by Carlos Garcés Manau, "Vincencio Juan de Lastanosa: Una Biografía," in *Vincencio Juan de Lastanosa (1607-1681): La Pasión de Saber*, 30.
59. Biblioteca Nacional de Madrid, Caja 18,727-45, "Las tres cosas mas singulares que tiene la Casa de Lastanosa en este año de 1639."
60. Burke and Cherry, *Collections of Paintings in Madrid*, xiii, 109-111, 116, 139-140; Brown, *Painting in Spain*, chap. 10.
61. Brown, *Painting in Spain*, 61; Jonathan Brown, "'Nos Quedamos Atonitos ante

la Cantidad de Pinturas': El coleccionismo regio en el siglo XVII," in Fernando Checa, ed., *El Real Alcázar de Madrid: Dos Siglos de Arquitectura y Coleccionismo en la Corte de los Reyes de España* (ex. cat. Madrid, 1994), 450.
62. Burke and Cherry, *Collections of Paintings in Madrid*, xiii, 40; Brown and Kagan, "The Duke of Alcalá: His Collection and its Evolution," 231-2; Sarah Schroth, "Early Collectors of Still-Life Painting in Castile," 28; Brown, *Painting in Spain*, 176; Vergara, *Rubens and his Spanish Patrons*, 185.
63. Jonathan Brown and John H. Elliott, *A Palace for a King: The Buen Retiro and the Court of Philip IV* (New Haven and London: Yale University Press, 1980), 118-121; quotation is from p. 121.
64. Brown, "'Nos Quedamos Atonitos ante la Cantidad de Pinturas,'" 449.
65. Kelley Helmstutler Di Dio, "The Chief and Perhaps Only Antiquarian in Spain: Pompeo Leoni and his Collection in Madrid," *Journal of the History of Collecting* 18 no. 2 (2006) 151, 155-6.
66. On the fashion for Rubens, see especially Alexander Vergara, *Rubens and his Spanish Patrons*.
67. HSA Ms. B-2424, ff. 52r-79v, "Narración de lo que le pasó a Don Vincencio Lastanosa a 15 de Octubre del Año 1662," ff. 76v-77r; HSA Ms. B-2424, ff. 24r-51v, Uztarroz, "Descripción del Palacio y Jardines de Vincencio Juan de Lastanosa," ff. 25r, 25v, 26v, 28r, 36r, 37v.
68. Brown and Kagan, "The Duke of Alcalá: His Collection and its Evolution," 242.
69. On the secular tastes of the great collectors, see Schroth, "Early Collectors of Still-Life Painting in Castile," 32-33; Brown, *Painting in Spain*, 2-3. On the taste for Flemish and Italian art, see Burke and Cherry, *Collections of Paintings in Madrid*, 2-6; Brown, *Painting in Spain*, 166, 173; Volk, "New Light on a Seventeenth-Century Collector"; Vergara, *Rubens and his Spanish Patrons*, 166; Fernando Checa and Jesús Sáenz de Miera, "La Corte Española y la Pintura de Flandes," in Checa, ed., *El Real Alcázar*, 220-235.
70. Morán and Checa, *El Coleccionismo en España*, 202.
71. On these features in gardens of the period, see Nieta Bedoya, "El espíritu y la esencia de la jardinería en los Países Bajos"; Lleó Cañal, "Los jardines de la nobleza"; Postma, "Een Zuidnederlandse Hovenier," 10; Harvey, "Spanish Gardens in their Historical Background," 9-11. On the form of Lastanosa's garden, see Rabanal Yus, "Los jardines de Vincencio Juan de Lastanosa"; Maria Celia Fontana Calvo, "En Torno a la Cronología y los Elementos del Jardín de Lastanosa," *Argensola* 115 (2005) 167-185.
72. Schroth, "Early Collectors of Still-Life Painting in Castile," 30.
73. López Terrada, "Las plantas ornamentales," 307, 319, 322, 325; Postma, "Een Zuidnederlandse Hovenier," 10, 17.
74. On Spanish collecting and social advancement, although here concerning an artist, Pompeo Leoni, see Di Dio, "The Chief and Perhaps Only Antiquarian in Spain," 151.
75. For example, Gracián's *El Politico* (Huesca: Juan Nognes, 1646), which sports Lastanosa's name on the title page and was dedicated to Philip IV, and, later,

Lastanosa's *Medallas Romanas Explicadas* (Huesca, 1675), dedicated to Juan José.
76. Carlos Garcés Manau, "Vincencio Juan de Lastanosa: Una Biografía," 37-8; Carlos Garcés Manau, "Un Lastanosa Poco Conocido (1665-1679): Las Relaciones con Juan José de Austria," *Argensola* 115 (2005) 41-93.
77. Vergara, *Rubens and his Spanish Patrons*, 182.

CONTRIBUTORS

Harold J. Cook is Director of the Wellcome Trust Centre for the History of Medicine and Professor at University College London (London, United Kingdom). His latest book *Matters of Exchange: Commerce, Medicine and Science in the Dutch Golden Age* (Yale University Press, 2007) is related to medicine and natural history in the Dutch Golden Age in an attempt to reassess the relationships between the beginnings of a world-wide trading system and a world-wide exchange of information about nature.

Antonio Barrera-Osorio (Ph.D. University of California, Davis, 1999) is Associate Professor of History at Colgate University, where he teaches the history of science in the sixteenth and seventeenth centuries, Atlantic World history (1500-1800), and early modern Spanish history. Some of his publications include *Experiencing Nature: The Spanish American Empire and the Early Scientific Revolution* (University of Texas Press, 2006); "Local Herbs, Global Medicines: Commerce, Knowledge, and Commodities in Spanish America," in *Merchants and Marvels: Commerce, Science, and Art in Early Modern Europe*, eds. Pamela Smith and Paula Findlen (New York: Routledge, 2001); and "Things from the New World: Commodities and Reports" in *Colonial Latin America Review* 15 (2006).

John Slater is Assistant Professor of Golden Age literature in the Department of Spanish and Portuguese at the University of Colorado, Boulder. His work focuses on the historiography and natural history of the Spanish Baroque. Some of his publications include "Fables of Communication: The Rhetoric of Investigative Methodology and Golden Age Literature." in *Beyond the Black Legend: Spain and the Scientific Revolution / Mas allá de la Leyenda Negra: España y la Revolución Científica*. Ed. Victor Navarro Brotòns and William Eamon. Valencia: Soler, 2007. He is currently working on a book that studies the ways in which early modern Spain's fascination with plants and flowers transformed the representation of power and patterns of literary discourse.

Jorge Cañizares-Esguerra is the Alice Drysdale Sheffield Professor of History at the University of Texas-Austin. He is author of two award-winning books: *How to Write the History of the New World* (2001) and *Puritan Conquistadors* (2006). He has also written *Nature, Empire, and Nation*, (2006) and co-edited with Erik Seeman of *The Atlantic in Global History, 1500-2000* (2006). He is currently studying historical narratives of Atlantic colonization understood as the fulfillment of prefigurations in four different types of texts (nature, the Bible, the classics, and indigenous sources).

Daniela Bleichmar is Assistant Professor of Art History and Spanish and Portuguese at University of Southern California. Her work focuses on the production and uses of visual material in science, the history of collecting and display, the history of the book, and the history of the Spanish empire. She is co-editor of *Science in the Spanish and Portuguese Empires (1500-1800)* (Stanford University Press, 2008).

Miguel López-Pérez (Ph.D. Universidad Complutense, Madrid, 2001). Independent Scholar. His research focuses on alchemy and medicine in Early Modern Spain. He is author of *Asclepio Renovado. Medicina y alquimia en la España Moderna, 1500-1700* (Madrid, Corona Borealis, 2003) and co-editor of *Los hijos de Hermes. Alquimia y espagiria en la terapéutica española moderna* (Madrid, Corona Borealis, 2001).

Maria M. Portuondo is an Assistant Professor at The Johns Hopkins University where she teaches early modern history of science and technology. She has a Bachelor of Science in electrical engineering and a Ph. D. in history of science and technology. She is the author of the *Secret Science: Spanish Cosmography and the New World* (University of Chicago Press, 2009).

William Eamon is Regents Professor of History and Dean of the Honors College at New Mexico State University. He is author of *Science and the Secrets of Nature: Books of Secrets in Medieval and Early Modern Europe* (Princeton University Press, 1994) and co-editor of *Beyond the Black Legend: Spain and the Scientific Revolution/Más allá de la Leyenda Negra: España y la Revolución Científica* (Valencia, Soler, 2007).

Bruce Moran is Professor of the history of science and medicine at the University of Nevada, Reno. He is author of *Distilling Knowledge: Alchemy, Chemistry, and the Scientific Revolution* (Harvard University

Press, 2005) and *Andreas Libavius and the Transformation of Alchemy: Separating Chemical Cultures with Polemical Fire* (Science History Publications, 2007).

Mar Rey-Bueno (Ph.D. Universidad Complutense, Madrid, 2000). Independent Scholar. Her research focuses on alchemy, medicine and magic in Early Modern Spain. She is author of *El hechizado. Medicina, alquimia y superstición en la corte de Carlos II* (Madrid, Corona Borealis, 1998) and *Los señores del fuego. Destiladores y espagíricos en la corte de los Austrias* (Madrid, Corona Borealis, 2002) and editor of *La Biblioteca Mágica* (Madrid, Universidad Complutense, 2007).

Rafael Chabrán is Professor of Spanish in the Department of Modern Languages and Literatures at Whittier College (Whittier, California). He has done extensive research and published on the life and works of Dr. Francisco Hernández and Miguel de Unamuno.

Anne Goldgar is Reader in Early Modern European History at King's College (London, United Kingdom). She specialises in seventeenth and eighteenth century European cultural and social history, especially of the Netherlands and of Francophone culture across Europe. She is also interested in print culture, the culture of collecting, and the interaction of society, art and science. Her latest book is about the ways in which tulipmania reflected deep anxieties about the transformation of Dutch society in the Golden Age and is entitled *Tulipmania: Money, Honor and Knowledge in the Dutch Golden Age* (University of Chicago Press, 2007).